S0-AKX-995

Applications of Scanned Probe Microscopy to Polymers

ACS SYMPOSIUM SERIES **897**

Applications of Scanned Probe Microscopy to Polymers

James D. Batteas, Editor
National Institute of Standards and Technology

Chris A. Michaels, Editor
National Institute of Standards and Technology

Gilbert C. Walker, Editor
University of Pittsburgh

Sponsored by the
ACS Division of Polymeric Chemistry: Science and Engineering, Inc

American Chemical Society, Washington, DC

Library of Congress Cataloging-in-Publication Data

Applications of scanned probe microscopy to polymers / James D. Batteas, editor, Chris A. Michaels, editor, Gilbert C. Walker, editor ; sponsored by the ACS Division of Polymeric Chemistry: Science and Engineering, Inc.

p. cm.—(ACS symposium series ; 897)

Includes bibliographical references and index.

ISBN 0–8412–3883–9 (alk. paper)

1. Polymers—Optical properties—Congresses. 2. Polymers—Microscopy—Congresses. 3. Scanning probe misroscopy—Congresses.

I. Batteas, James D. II. Michaels, Chris A. III. Walker, Gilbert C. IV. American Chemical Society. Division of Polymeric Materials: Science and Engineering, Inc. V. Series.

QD381.9.O66A64 2005
620.1′9204295—dc22 2004062662

The paper used in this publication meets the minimum requirements of American National Standard for Information Sciences—Permanence of Paper for Printed Library Materials, ANSI Z39.48–1984.

Distributed by Oxford University Press

PRINTED IN THE UNITED STATES OF AMERICA

Foreword

The ACS Symposium Series was first published in 1974 to provide a mechanism for publishing symposia quickly in book form. The purpose of the series is to publish timely, comprehensive books developed from ACS sponsored symposia based on current scientific research. Occasionally, books are developed from symposia sponsored by other organizations when the topic is of keen interest to the chemistry audience.

Before agreeing to publish a book, the proposed table of contents is reviewed for appropriate and comprehensive coverage and for interest to the audience. Some papers may be excluded to better focus the book; others may be added to provide comprehensiveness. When appropriate, overview or introductory chapters are added. Drafts of chapters are peer-reviewed prior to final acceptance or rejection, and manuscripts are prepared in camera-ready format.

As a rule, only original research papers and original review papers are included in the volumes. Verbatim reproductions of previously published papers are not accepted.

ACS Books Department

Contents

Near-Field Scanning Optical Microscopy

Mechanical Studies of Polymers by Atomic Force Microscopy

Single-Molecule Studies

Polymer Surface Characterization by Scanned Probes

Indexes

Preface

Since its inception in 1981, scanning probe microscopy has developed into an essential tool for the characterization of surfaces and interfaces. Whether it is simply measurements of the often unexpected and dramatic structures of surfaces at the nanoscale or the examination of the properties of single molecules, advances in probe microscopies continue to reveal ever more detail of the nature of materials. The exploration of polymers, either natural (such as proteins) or synthetic, has benefited significantly from these advances. Scanned probes offer the unique ability for in situ probing of nanoscale structure and mechanical properties, which are critical aspects of many polymer systems. It is the capability of probe microscopies, and in particular atomic force microscopy (AFM), to interrogate surfaces and interfaces under physiological conditions that has opened the door to biophysical studies of the fundamental nature of proteins and other biologically relevant systems at the single-molecule level. Improvements in rapid scanning and microscope design are offering expanded capabilities for dynamic measurements to allow time-lapsed views of changing polymer land-scapes. The exploitation of AFM for direct measurements of surface mechanical properties also shows significant promise for improvements in thin film design. Lastly, the integration of optical spectroscopies with the high spatial resolution imaging capabilities of scanned probes affords the concomitant interrogation of local structure and chemistry at the nanoscale. Many more advancements are sure to come in the near future making scanned probes evermore useful for studies of polymers and biopolymers.

This book draws together examples of the applications of scanning probe microscopies to natural and synthetic polymers to reveal dramatic new details of the nature of our world at the nanoscale. In the first part of this book, combined optical and structural studies utilizing near-field scanning optical microscopy illustrate the power of how combined local probes of chemical and structural information can help us to obtain clear pictures of polymer electrical properties (Chapter 1), polymer thin films (Chapter 2), polymer liquid crystals (Chapter 3), spatial chemical mapping of polymers in the infrared (Chapters 4 and 5), and the optical properties of semicrystalline polymers (Chapter 6).

In the second part of this book, the ability of AFM to probe changes in local mechanical properties, including methods for extracting mechanical information from AFM measurements (Chapters 7 and 10), applications of these measurements to biopolymers (Chapter 9), as well as mapping mechanical changes in confined polymer films (Chapter 8), are explored. These measurements of polymer mechanical properties are extended to the extreme limit of single polymer molecules in the third section of the book (Chapters 11 to 13).

We conclude this book with several examples of scanned probe character-ization of the structural properties of polymers, including measurements of dynamic processes (Chapters 14 and 15), industrial studies of polymers by AFM (Chapter 16), as well as how the detailed organization of blended polymers can be investigated using AFM (Chapter 17). We hope that this collection of examples helps to both illustrate the power of scanned probes for the investigation of polymer surfaces and to provide a hint of what is to come in the development of this ever evolving measurement technology.

Acknowledgments

The symposium from which this book was drawn was held at the 214th National American Chemical Society (ACS) Meeting in New Orleans, Louisiana in March of 2003. We gratefully acknowledge support for the symposium from the ACS Division of Polymeric Materials: Science and Engineering, Inc. (PMSE), the Petroleum Research Fund (PRF# 39350–SE), the Chemical Science and Technology Laboratory of the National Institute of Standards and Technology, and RHK Technology,

Inc. The editors also gratefully acknowledge the reviewers and authors for their excellent contributions to this book.

James D. Batteas
National Insitute of Standards and Technology
100 Bureau Drive, Stop 8372
Gaithersburg, MD 20899
(301) 975–8907 (telephone)
(301) 926–6689 (fax)
james.batteas@nist.gov (email)

Chris A. Michaels
National Insitute of Standards and Technology
100 Bureau Drive, Stop 8372
Gaithersburg, MD 20899
(301) 975–5418 (telephone)
(301) 926–6689 (fax)
chris.michaels@nist.gov (email)

Gilbert C. Walker
Department of Chemistry
Chevron Science Center
University of Pittsburgh
219 Parkman Avenue
Pittsburgh, PA 15260
(412) 383–9650 (telephone)
(412) 383–9646 (fax)
gilbertw+@pitt.edu (email)

Near-Field Scanning Optical Microscopy

Chapter 1

Near-Field Spectroscopic Studies of Fluorescence Quenching by Charge Carriers

Andre J. Gesquiere, Doo Young Kim, So-Jung Park, and Paul F. Barbara*

Center for Nano-and Molecular Science and Technology, Department of Chemistry and Biochemistry, University of Texas, Austin, TX 78712
***Corresponding author: p.barbara@mail.utexas.edu**

Near-field scanning optical microscopy (NSOM) with an electrically biased probe is a technique that allows the imaging of charge carrier drift (mobility), carrier concentration, and their interaction with excitations (excitons) in functioning devices. This technique has been applied to poly[2-methoxy, 5-(2′-ethyl-hexyloxy)-*p*-phenylenevinylene] (MEH-PPV) and tetracene-doped pentacene devices. Space and time resolved data show that quenching by charge carriers and charge trapping by defect sites in the organic materials, specifically in the presence of oxygen, is found to be of critical importance in the understanding and further development of more efficient and durable devices.

Introduction

Organic thin films are being researched for and applied in a number of innovative technologies such as light emitting displays, solar cells and sensors. The morphology of these organic thin films is known to be a critical factor in device function and performance. Several techniques are in place that allow for direct imaging of thin film morphology, from the macroscopic down to the nanoscopic scale. At the nanoscale, atomic force microscopy (AFM), scanning electron microscopy (SEM) and near-field scanning optical microscopy (NSOM) (*1,2,3,4*) have become "household instruments" for determining thin film morphology.

NSOM is a scanning probe optical microscopy that surpasses the diffraction limit (λ/2) inherent to conventional far field microscopy, thus achieving unrivaled optical resolution i.e. down to 20-30 nanometer (nm). This is achieved by bringing an optical fiber with an aperture much smaller than the wavelength of the light within a distance of a sample that is ~5 nm. The combination of a sub-wavelength sized point source of light brought within a few nanometer of a sample surface results in tremendous optical resolution. Since its inception (*1,2,3,4*) NSOM has been extensively utilized to determine the nanostructure of a variety of materials, most importantly of complex organic thin film materials.

Figure 1 depicts a typical NSOM setup. Laser light is coupled into an optical fiber, which has been pulled to a narrow point at the end leaving a nanometer sized aperture. This end is usually coated with aluminum to force the light through the aperture.(*5,6*) The probe sample distance is typically regulated through a tuning fork based shear-force feedback scheme.(*7,8,9*) This approach provides a highly sensitive mechanism for tip-sample control, and eliminates the need for the more cumbersome shear-force feedback mechanism in which an additional laser and photodiode are required. (*5,10,11*)

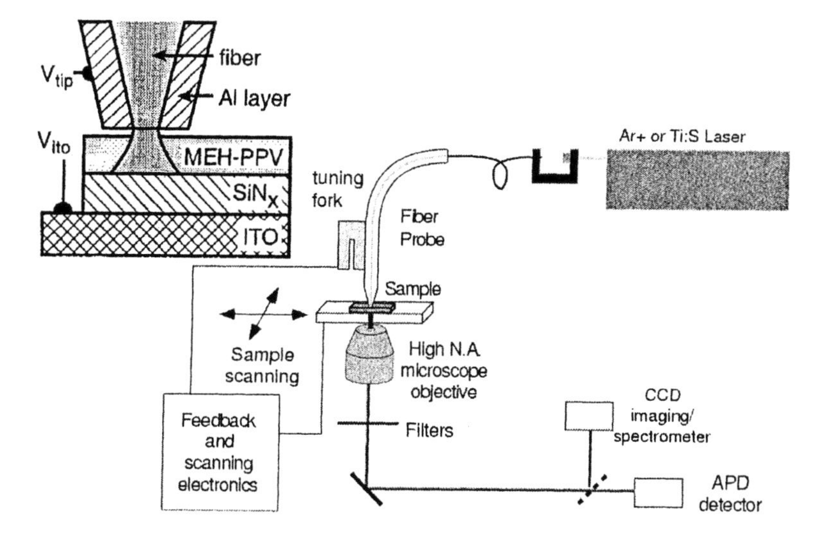

Figure 1: Scheme of a typical NSOM setup based on a tuning fork feedback mechanism. The inset shows a sketch of the thin film MEH-PPV sample with the electrically biased probe nearby. (Adapted from reference 19. Copyright 2001 American Chemical Society)

During the experiments the sample is raster scanned in the (x,y)-plane. For each pixel in the image the topography and fluorescence/transmission signal is simultaneously recorded. The fluorescence and transmission images differ in having spectral filtering in place or not, respectively. Besides imaging, chemical contrast can be achieved by localizing the tip over the sample and acquiring spectra of the sample.

Even though optical resolution below 50nm can be achieved, it is quite uncommon. The required aperture size prevents sufficient amounts from leaking out of the fiber towards the sample.

This chapter deals with charge diffusion in thin films and single crystals, and in particular the interaction between hole polarons and excited states in conjugated polymer thin films. The first section discusses a nanoscale study of MEH-PPV (poly[2-methoxy, 5-(2′-ethyl-hexyloxy)-*p*-phenylenevinylene]) thin film fluorescence modulated by an electrical bias applied between the NSOM probe and the sample under investigation. MEH-PPV is one of the most researched materials in view of its application in organic light emitting diodes (OLEDs). Properties such as solubility, brightness and easily accessible energy levels have contributed to the widespread acceptance of MEH-PPV as an excellent material for OLED fabrication. In a second section of this chapter a similar study as described in section one is briefly discussed. Single crystals of tetracene and its blends with pentacene, both small molecules, have been studied with NSOM using an electrically biased tip. These studies reveal comparable phenomenological observations as determined for the conjugated polymer MEH-PPV, however with a strongly different origin.

Field-Induced Photoluminescence Modulation of MEH-PPV under Near-Field Optical Excitation

Conjugated polymers have raised considerable interest for several years due to their possible application in light emitting displays.(*12,13*) Consequently, the electrooptical properties of these materials are continuously under investigation, since the optical and electrical properties of these materials still prove to be a challenging hurdle in achieving the goal of device commercialization. The main problem with OLED devices is the existence of trap states in the device. The density and energy distribution (deep trap vs shallow trap) of these traps influence device properties such as charge balance and device turn-on voltage.(*14*) In a functioning conjugated polymer based OLED device the charge carriers are believed to be localized polarons.(*14*) Excitations in the conjugated polymer material (excitons, which are neutral) are localized as well.

It is known that conjugated polymer thin films are nanostructured.(*15,16*) This nanostructure strongly influences the device performance and properties such as operating voltages and charge carrier mobilities. Control of morphology has been shown to improve device function.(*17*) Indeed, the physical processes related to device function such as energy and charge transport, exciton dissociation and charge injection are subject to molecular scale interactions. A technique has been developed by our group to study space resolved field-induced fluorescence modulation.(*18*) In this technique, based on NSOM, an electrical bias is applied between the NSOM probe and the sample, making it possible to map organic device function with sub-100 nanometer resolution.

The sample used in this study (see Figure 1) consists of an MEH-PPV thin film spin coated on top of an optically transparent indium-tin oxide (ITO) electrode coated with a silicon nitride (SiN_x) blocking layer.(*19,20*) The SiN_x layer prevents charge injection from the ITO electrode into the MEH-PPV film.

In Figure 2 a topography and fluorescence image, acquired simultaneously for the same region of the sample, are presented. The topography of the MEH-PPV film is related to the roughness of the ITO substrate surface. In the fluorescence image that is directly correlated with this topography image one can see that the fluorescence intensity is lower for raised topographical features. The higher features in the topography image thus correspond with a thinner MEH-PPV film.

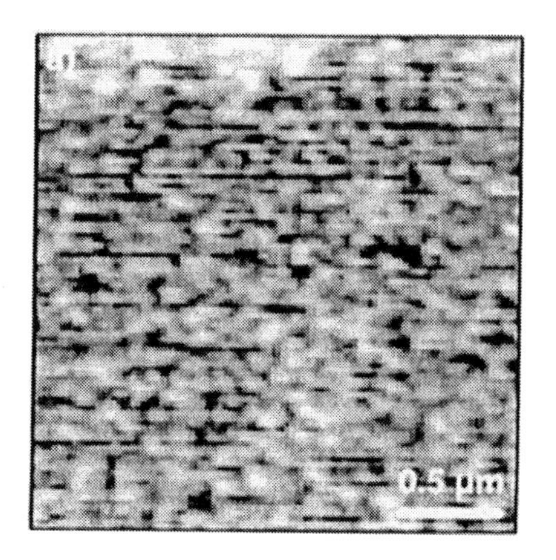

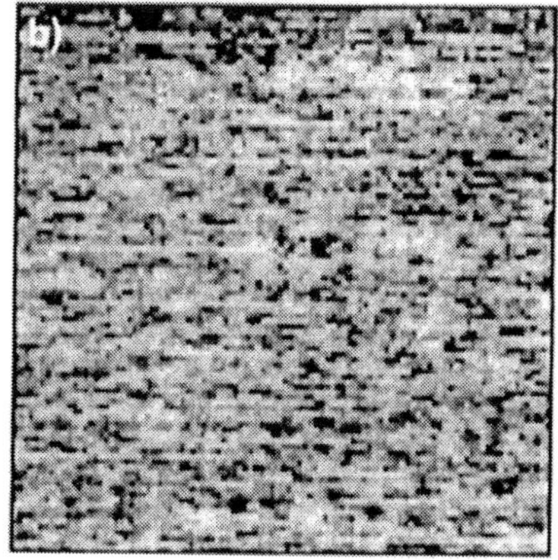

Figure 2: NSOM images of a MEH-PPV film on ITO. Topography (a) and correlated fluorescence image (b) are shown. (Reproduced from reference 19. Copyright 2001 American Chemical Society)

By wiring the tip and sample to a function generator voltage sequences could be applied to the sample. A typical example is shown in Figure 3. The top panel shows the applied voltage sequence, the bottom panel shows the corresponding fluorescence modulation averaged over many cycles to achieve an adequate signal to noise ratio. The data shows that with the tip biased positive (sample negative) the photoluminescence intensity is enhanced, while the photoluminescence intensity is reduced with a negative biased probe (sample positive). Fluorescence quenching due to exciton dissociation has been observed for samples employing a planar electrode geometry. However, in those cases fluorescence enhancement was never observed. The modulation of the fluorescence intensity with applied bias observed in the NSOM experiments is consistent with charge carrier induced quenching and a field-induced modulation of the charge carrier concentration in the area under the tip. The fluorescence modulation depth decreases linearly with tip-sample separation (E=V/d), confirming that the observed phenomena are indeed a field effect.

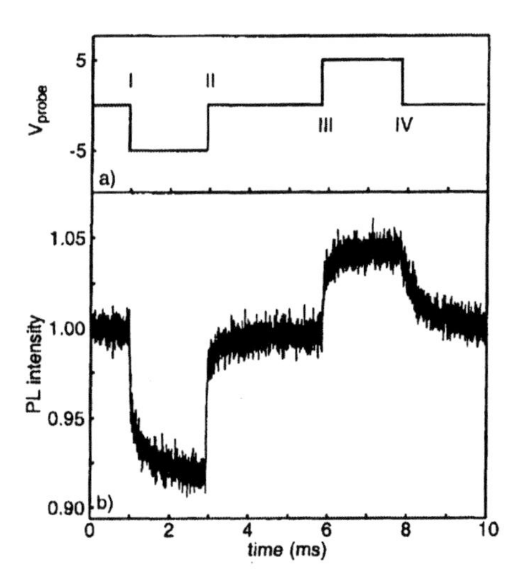

Figure 3: MEH-PPV fluorescence time trajectory acquired while applying the voltage sequence shown in the top panel. The data is averaged over many cycles. (Reproduced from reference 20. Copyright 2002 American Chemical Society)

The data in Figure 3 also contains information on the kinetics associated with the field-induced fluorescence modulation caused by a modulation of the charge carrier concentration under the area of the NSOM tip. After turning on a -5V bias on the probe (I in Figure 3), the fluorescence intensity drops in a non-instantaneous manner. A fast drop in intensity is followed by a slow relaxation to a lower intensity level. When the negative bias is removed (II in Figure 3) the fluorescence recovers with a fast and slow component in the process as well.

The increase in fluorescence intensity when applying a positive bias on the NSOM probe shows an analogous behavior (III and IV in Figure 3). After an initial fast change in intensity a slow relaxation occurs. The observed temporal response indicates the presence of trapped carriers in the device that have low mobility, and is consistent with the drift of hole polarons towards and away from the area under the NSOM tip.

The presence of traps for charge carriers in OLEDs has been related to the presence of oxygen on several occasions. To investigate the effect of oxygen on the transient response of the fluorescence signal to the applied bias experiments were carried out both in ambient air and under a dry nitrogen atmosphere.(*20*) An airtight enclosure built around the NSOM scanning unit was purged for several hours with dry nitrogen gas in order to rigorously remove oxygen from the experimental environment. The field-induced fluorescence modulation depth is larger when the sample is studied in ambient air (Figure 4). An increase in the amplitude and time constant of the slow component can also be observed under ambient conditions. This fits well with the notion that the presence of oxygen results in a higher density of trap states in the MEH-PPV film.

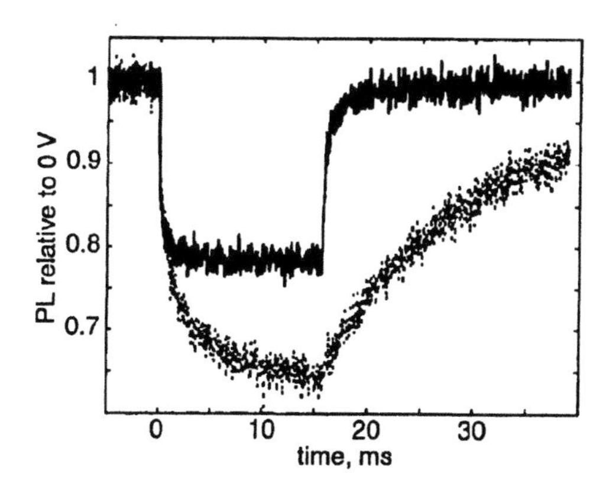

Figure 4: Averaged MEH-PPV time transients acquired in ambient air (dotted line) and nitrogen atmosphere (solid line). A -5V pulse was applied. (Reproduced from reference 20. Copyright 2002 American Chemical Society)

In addition, O_2^- ions with low mobility might contribute to the observed slow component in the transient response of the fluorescence signal. Molecularly dissolved oxygen in conjugated polymer films is known to act as an electron acceptor, forming the superoxide ion (O_2^-) while p-doping the conjugated polymer.(*21*) In air the photoluminescence intensity of the MEH-PPV thin films observed with NSOM is 40% lower than in a nitrogen atmosphere. This photoluminescence decrease is caused by photooxidation (*22,23*) of the polymer in the presence of oxygen, increased photogeneration of charge carriers at defects created by photooxidation (i.e. carbonyl defects) and quenching by the formation of an MEH-PPV/ O_2 charge transfer complex.(*24,25,26,27*)

Field-Induced Photoluminescence Quenching by Charge Carriers in Pentacene-Doped Tetracene

The NSOM technique with an electrically biased probe has also been applied in the study of thin films of pentacene-doped tetracene.(*28*) These blends serve as a model system for organic materials with a low density of low energy trap sites. Single crystals of tetracene were mounted on epoxy coated ITO glass (Figure 5A).

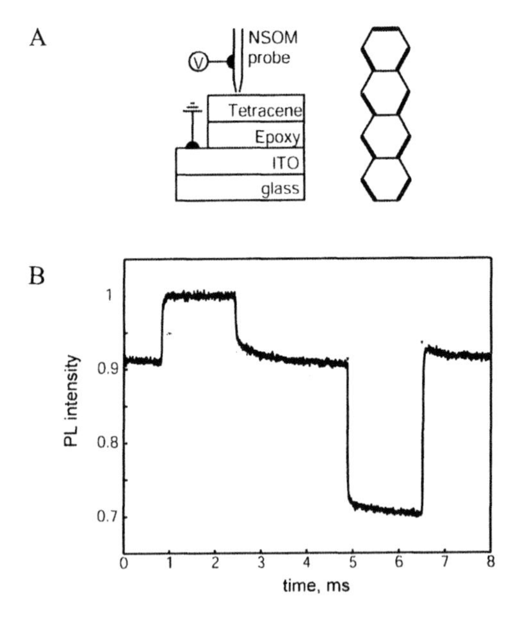

Figure 5: A) Scheme of the sample used in this study. An electrical bias is applied on the NSOM probe. B) Field-induced fluorescence modulation transient acquired for a pentacene-doped tetracene sample. (Adapted from reference 28)

Tip and sample were electrically biased by means of a function generator which applies the desired voltage sequence. When tetracene is optically excited with the NSOM tip energy is transferred from tetracene to pentacene and pentacene emission is observed. Applying a bias on the NSOM probe results in field-induced fluorescence modulation, as can be seen in Figure 5B. These data are analogous to the results for MEH-PPV thin films described in the previous section.(*19,20*) Again a positive biased probe results in a fluorescence intensity enhancement, while a negative biased probe induces a decrease in fluorescence intensity. Slow dynamics as those observed for MEH-PPV can be observed. Current-voltage experiments on organic devices have shown that these slow processes are related to the presence of deep hole traps, which are common in organic thin film devices.(*29,30,31,32*)

The fluorescence increase with the NSOM probe at positive bias is attributed to repulsion of hole polarons from the area under the NSOM tip, and depends on the dopant concentration. At negative bias, hole polarons are attracted towards the area under the tip, quenching the pentacene emission. The quenching ratio decreases with increasing dopant concentration, supporting the notion that pentacene sites are filled with holes. Emission spectra (not shown) strongly suggest that hole polarons are trapped by pentacene. So, hole polarons reside in the pentacene sites, and their density is determined by the dopant density. However, there is no simultaneous increase in tetracene fluorescence intensity. This indicates that pentacene cations quench tetracene fluorescence at a rate comparable to that of neutral pentacene. As evident from these data, both charge transfer and energy transfer to low energy trap sites in organic materials are important factors to be considered in device fabrication.

Summary

Near-field optical spectroscopy with electrically biased probes is a useful technique for imaging device function and charge carrier drift. Spatially and time-resolved information on charging phenomena can be obtained at the nanometer scale. A broad array of materials such as for example nanostructured polymers, polymer blends, organic material-inorganic nanoparticle composites and single crystals can be studied with this technique in a device-like geometry to improve the understanding of local fields, carrier concentrations and mobilities in these materials.

Acknowledgements

We gratefully acknowledge the National Science Foundation and the Welch Foundation for support of this research.

References

1. Pohl DW. 1991. Advances in Optical and Electron Microscopy, ed. T Mulvey, CJR Sheppard, 12:243–312. New York: Academic.
2. Betzig, E.; Trautman, J. K. *Science* **1992**, *257,* 189.
3. Kopelman, R.; Tan, W. H. *Appl. Spec. Rev.* **1994**, *29*, 39.
4. Heinzelmann, H.; Pohl, D. W. *Appl. Phys. A* **1994**, *59*, 89.
5. Betzig, E.; Trautman, J. K.; Harris, T. D.;Weiner, J. S.; Kostelak, L. *Science* **1991**, *251*, 1468.
6. Betzig, E.; Grubb, S. G.; Chichester, R. J.; Digiovanni, D. J.; Weiner, J. S. *Appl. Phys. Lett.* **1993**, *63*, 3550.
7. Karrai, K.; Grober, R. D. *Appl. Phys. Lett.* **1995**, *66*, 1842.
8. Karrai, K.; Grober, R. D. *Ultramicroscopy* **1995**, *61*, 197.
9. Ruiter, A. G. T.; Vanderwerf, K. O.; Veerman, J. A.; Garciaparajo, M. F.; Rensen, W. H. J.; Vanhulst, N. F. *Ultramicroscopy* **1998**, *71*, 149.
10. Betzig, E.; Finn, P. L.; Weiner, J. S. *Appl. Phys. Lett.* **1992**, *60*, 2484.
11. Toledo-Crow, R.; Yang, P. C.; Chen, Y.; Vaez-Iravani, M. *Appl. Phys. Lett.* **1992**, *60*, 2957.
12. Burroughes, J. H.; Bradley, D. D. C.; Brown, A. R.; Marks, R. N.; MacKey, K.; Friend, R. H.; Burn, P. L.; Holmes, A. B. *Nature* **1990**, *347*, 539.
13. Braun, D.; Heeger, A. J. *Appl. Phys. Lett.* **1991**, *58*, 1982.
14. Antoniadis, H.; Abkowitz, M. A.; Hsieh, B. R. *Appl. Phys. Lett.* **1994**, *65*, 2030.
15. Barbara, P.; Adams, D.; O'Connor, D. *Annu. Rev. Mater. Sci.* **1999**, *29*, 433.
16. Winokur, M. J. In Handbook of Conducting Polymers, 2nd ed.; Skotheim, T., Elsenbaumer, R., Reynolds, J., Eds.; Marcel Dekker: New York, 1998.
17. Nguyen, T.-Q.; Kwong, R. C.; Thompson, M. E.; Schwartz, B. J. *Appl. Phys. Lett.* **2000**, *76*, 2454.
18. Adams, D. M.; Kerimo, J.; Liu, C. Y.; Bard, A. J.; Barbara, P. F. *J. Phys. Chem. B* **2000**, *104*, 6728.
19. McNeill, J. D.; O'Connor, D. B.; Adams, D. M.; Kämmer, S. B.; Barbara, P. F. *J. Phys. Chem. B* **2001**, *105*, 76.
20. McNeill, J. D.; Barbara, P. F. *J. Phys. Chem. B* **2002**, *106*, 4632.
21. Harrison, M. G.; Gruner, J.; Spencer, G. C. W. *Phys. Rev. B* **1997**, *55*, 7831.
22. Antoniadis, H.; Rothberg, L. J.; Papadimitrakopolous, F.; Yan, M.; Galvin, M. E.; Abkowitz, M. A. *Phys. Rev. B* **1994**, *50*, 14911.
23. Yan, M.; Rothberg, L. J.; Papadimitrakopolous, F.; Galvin, M. E.; Miller, T. M. *Phys. Rev. Lett.* **1994**, *73*, 744.
24. Abdou, M. S. A.; Orfino, F. P.; Son, Y.; Holdcroft, S. *J. Am. Chem. Soc.* **1997**, *119*, 4518.
25. Yu, J.; Hu, D. H.; Barbara, P. F. *Science* **2000**, *289*, 1327.

26. Park, S.-J.; Gesquiere, A. J.; Yu, J.; Barbara, P. F. *J Am. Chem. Soc.* **2004**, in press.
27. Gesquiere, A. J.; Park, S.-J.; Barbara, P. F. *J. Phys. Chem. B* **2004**, in press.
28. McNeill, J. D.; Kim, D. Y.; Yu, Z.; O'Connor, D. B.; Barbara, P. F. *J Phys Chem B* **2004**, submitted.
29. Bozano, L.; Carter, S. A.; Scott, J. C.; Malliaras, G. G.; Brock, P. J. *Appl. Phys. Lett.* **1999**, *74*, 1132.
30. Berleb, S.; Brutting, W.; Paasch, G. *Organic Electronics* **2000**, *1*, 41.
31. Berleb, S.; Brutting, W.; Paasch, G. *Synthetic Metals* **2001**, *122*, 37.
32. Paasch, G.; Scheinert, S. *Synthetic Metals* **2001**, *122*, 145.

Chapter 2

Time-Resolved Fluorescence Near-Field Scanning Optical Microscopy Studies of Conjugated Polymer Thin Films

Joseph M. Imhof, Eun-Soo Kwak, and David A. Vanden Bout

Department of Chemistry and Biochemistry, Center for Nano- and Molecular Science and Technology, and Texas Materials Institute, University of Texas, Austin, TX 78712

Fluorescence lifetime measurements have been coupled with high resolution optical imaging by time-correlating single photons in a near-field scanning optical microscope. The technique provides a useful tool to study the heterogeneity of polymer thin films. The fluorescence lifetime provides a significantly more robust measure of the fluorescent properties of conjugated polymer thin films than the simple fluorescence as it is less subject to scanning artifacts that alter the total fluorescence images. The technique has been used to study the effects of small insoluble cluster and polymer chain order on the luminescence of poly(dialkylfluorene) films as well as heterogeneities and quenching by metal electrodes in poly(phenyleneethyenelene) films. The results show that all of the films are extremely uniform despite what might be observed in total fluorescence images. The only notable lifetime features result from deliberate photochemistry or quenching by an added metal layer.

Introduction

There has been a great deal of interest in the optoelectronic properties of conjugated polymers due to high photo and electroluminescence character that makes them extremely promising in light emitting devices.[1] These polymers can be easily processed into thin films by spin casting onto a substrate forming highly ordered crystalline structures with small domains. The film spectroscopy is more complicated than that of the polymers in solution. Characterization of the emission on small distance scales is critical in determining how nanoscale heterogeneities contribute to the emission species observed in the films. Stiff chain polymers like those of the poly (p-phenylene), poly (p-phenyleneethnylene) and polyfluorene families have been studied and utilized in LED displays and semi conducting materials due to their high anisotropy properties accessed via processing through the liquid crystalline state.[2-4] Anisotropic nature is introduced by different conditions of packing of the polymer. However, the packing of thin polymer films through the liquid crystalline state is not well understood and thus microscopic measurements of polymer ordering are needed to control and understand this process. Bulk measurements of molecular orientation of these polymers may hide important sample details since these measurements average across a specific distribution of individual molecules. NSOM is a great tool for obtaining fluorescence information while simultaneously collecting topographic data with high spatial resolution.[5] However, the fluorescence intensity in such images is often altered due to scattering, self absorption, tip-sample fluctuation, and scanning artifacts. The fluorescence lifetime of an emitting species however, will show little or no effect from these complicating factors. Fluorescence Lifetime Imaging can be combined with NSOM (FLI-NSOM) to yield further details concerning the heterogeneities in these polymer films.[6,7]

FLI-NSOM can also be utilized to study luminescence quenching that can result from contact between a light emitting polymer and a metal electrode. It has been repeatedly demonstrated that nonradiative quenching of an excited state fluorophore occurs near a metal surface.[8] The metal surface, when in close proximity to a fluorescent species, acts as an energy sink to provide a nonradiative route for fluorophore relaxation to occur from the first singlet excited state back to the ground state. The simplest organic light emitting device (OLED) is composed of three layers. A transparent conducting electrode such as indium tin oxide (ITO) is coated with a 50-100 nm thick layer of light emitting material (in this case a polymer). A metal cathode is vapor deposited over the organic layer to complete the circuit and yield a finished OLED. The concern with these devices is that the small dimensions of the organic active layer leaves excited state fluorophores especially susceptible to nonradiative quenching near the conductive interfaces of the sandwich structure. We will investigate this interface by modelling a typical sandwich OLED structure with a planar geometry. A polymer/metal interface will be fabricated using gold and the conjugated polymer, di-dodecyl poly(phenyleneethynylene) (DPPE). NSOM imaging of pristine film over glass and over the metal allows for a direct comparison of the lifetime of these regions.

Experimental

Sample Preparation

Preparation of the polymer and polymer/metal interface specimens used to investigate quenching of a conjugated system near a metal boundary were prepared as follows. Microscope cover slip substrates were soaked in a basic aqueous solution (15% KOH) under sonication for 10 minutes to yield clean, uniform substrate surfaces. The cover slips were then triply rinsed with deionized water and sonicated in a pure water bath for an additional 10 minutes before being dried in a 70°C oven. For samples with gold films, the cleaned substrates were then coated with gold (100 nm) via sputter coating to yield smooth, uniform metal thin films. Each gold coated cover slip was then scored with a sharp razor edge to leave a smooth, crisp, linear metal boundary. The polymer films were deposited from solution either on clean cover-glass or gold coated cover slips by spin casting. Poly(dihexylfluorene) (PFH) films were cast from 2wt % stock solution in toluene to form 60 -100 nm thick film. Annealed polyfluorene samples were made by heating the pristine films in a dry nitrogen atmosphere for at 200°C for 1 hour. DPPE films were cast from a 0.5 wt% chloroform solution. The resulting 45 nm thick polymer film provided uniform coverage of the substrate surface with little roughness across the metal/glass threshold. This method provides an easily located and identified metal/polymer boundary for further NSOM analyses.

FLI-NSOM Instrumentation

Fluorescence lifetime imaging (FLI) near-field scanning optical microscopy (NSOM) is a technique that allows for the simultaneous acquisition of highly spatially resolved images with accompanying time resolved fluorescence data. These measurements are made by coupling the high spatial resolution of NSOM with the time resolved benefits of time-correlated single photon counting (TCSPC) technology. A commercial NSOM instrument (Aurora, Thermomicroscopes/Digital Instruments) has been modified to acquire both fluorescence and polarization measurements.[9] The microscope employs a frequency doubled Ti-Sapphire laser emission as a 400 nm excitation source for fluorescence measurements. The laser produces a train of approximately 100 fs pulses at a repitition rate of 81 MHz. The pulse train is coupled into a fiber optic NSOM probe fabricated in-house.[10] The sample is held in the near-field of the NSOM probe using a tuning fork detected shear-force mechanism.[11] Sample fluorescence is collected using a high N.A. microscope objective and a series of long pass filters. The fluorescence signal is split into two orthogonally polarized components. One component is detected with an avalanche photodiode single-photon counting module (APD) while the other orientation is collected via a multichannel plate detector (MCP) that is used for fluorescence decay acquisition. The APD is used for real-time fluorescence monitoring at every image pixel, while the MCP collects lifetime decays at every point to be analyzed later. Both the time-to-amplitude converter and multi channel analyzer required for TCSPC measurements are contained on a single PCI card (Timeharp 1000). In this way, an entire fluorescence decay can be collected at every pixel via TCSPC. The decays are recorded in reverse timing mode.[12] The time resolution of the current instrument is limited by the response of the employed MCP detector. Instrument response of the system was calculated by examining the decay profile exhibited by the MCP of the transmitted laser pulse train through an uncoatd glass microscope coverslip. The measured instrument response was 500 ps.

Results and discussion

Polyfluorene films

Collection of the full time correlated lifetime is not only useful for image contrast but also allows decays of interested regions to be created. Since regions of interest can easily be identified in the images, averaged decays can be measured for each of these regions by binning together the decays of these selected pixels. Previous studies showed that pristine poly (9,9′ – dihexylfluorene) films have clusters from the insolubility of the polymer in the spinning solvent prior to spin casting.[13] Clusters exhibit a decrease in total fluorescence that can be seen in the dark spots that correlate with each polymer cluster in the topography image as can be seen in Figure 1.

Figure 1 shows time-resolved NSOM images of pristine poly (9,9′ – dihexylfluorene) film. The film has been imaged in 2 x 2 μm first and then

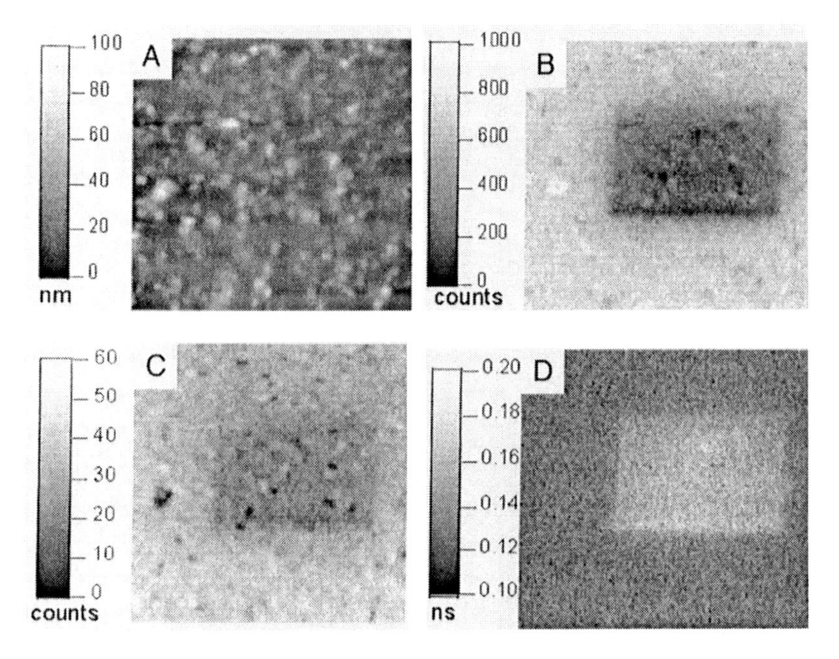

Figure 1: Time-resolved fluorescence NSOM images of pristine poly (9,9′– dihexylfluorene) thin film.The image is 5 x 5 μm. The regions of contrast in the center isfrom a previous 2 x 2 μ scan.. (A) is the topography, (B) is the total fluorescence, (C) is the total fluorescence on MCP detector which is orthogonal polarization to (B), and (D) is lifetime calculated by using equation 1

imaged at 5 x 5 μm. Figure 1A shows film topography, indicating 50 – 150 nm sizes of clusters with heights ranging from 25 to a 100 nm. Figure 1B is an image of the total fluorescence signal on the APD and the clusters have reduced fluorescence. The previously scanned area is evident as the dark box due to less fluorescence signal from photobleaching. Figure 1C is the total fluorescence collected with the MCP detector. This signal is orthogonally polarized to the APD and the clusters again show dark spots with low counts. Figure 1B and 1C are two images taken simultaneously at orthogonal polarizations and they show identical regions of cluster correlated fluorescence. As in the previous study, these clusters are likely formed due to the insolubility of the polymer in the solvent and result in amorphous polymer chunks with isotropic fluorescence. A fluorescence lifetime image can be calculated by measuring the ratio of photons emitted at long time delays to the total number of photons emitted at each pixel. The sample lifetime can be estimated using,

$$\langle \tau \rangle = -T \Big/ \ln\left(I_f \Big/ I_T \right) \tag{1}$$

where I_T is the total fluorescence intensity and I_f is the fraction of the total intensity after a delay time T.[6] In the case of a single exponential decay equation 1 yields an exact lifetime, for multiexponential decays it will yield and average lifetime.[6] Figure 1D shows the lifetime calculated using equation 1 with T = 0.5 ns. The clusters, which have less total fluorescence than the remainder of the film, show the exact same lifetime as the rest of the film. This is a result of either the reduced fluorescence being an artifact from the changing topography on top of the clusters or the quenching being so fast that it is beyond the time-resolution of the current setup. The region in the center of the films that has been previously scanned shows a clear increase in the lifetime. This region has been purposely photobleached to yield contrast in the lifetime image. The increase lifetime has been previously observed[6] and is attributed to the generation of fluoronone defects along the polyfluorene chain.[14] The fluorescence decay of the PFO is extremely fast and the decays measured are nearly instrument limited. None the less it is clear that the films yields a uniform lifetime with the exception of the region with photochemistry. While the shot noise from the data will limit the accuracy of the lifetime to ±0.1 ns, this is the uncertainy of any given pixel. The homogeneity is revealed in the uniforminty of the signal over many pixels.

Images of thermally annealed films of poly (9,9′ – dihexylfluorene) show strong polarization contrast. The contrast is due to polymer ordering via a liquid crystalline phase transition upon annealing. A polarizing beam splitter is used to divide the fluorescence into two orthogonal polarizations that can be collected on separate detectors, in this case, an APD and a MCP. Two orthogonal images are taken simultaneously in order to obtain fluorescence images and lifetime images. Figure 2A is the shear force topography measured by raster scanning of the piezo over the film and 2C and 2D are the fluorescence images collected at orthogonal polarizations (Figure 2D shows the fluorescence at long time delays on the MCP). The size and shape of the domains result from a convolution of the tip with the polymer features. The domains are 30 – 60 nm wide and 100 – 150 nm long. Using NSOM these few tens of nanometer-sized domains can be revealed. In pristine film images, there are two lifetime components, one with a decay constant of 200 ps and and second with a long lifetime > 7ns.[15] Thermally annealed films show increased longer emission and a higher concentration of the long lifetime species. As the fluoronone defects are only observed in the films and not in solution it is possible they result from interchain interaction and may have an emission that is polarized perpendicular to the backbone of the polymer chains. If the two emitting species with different lifetimes emitting from states that are orthogonally polarized, the emission from aligned domains in 2C and

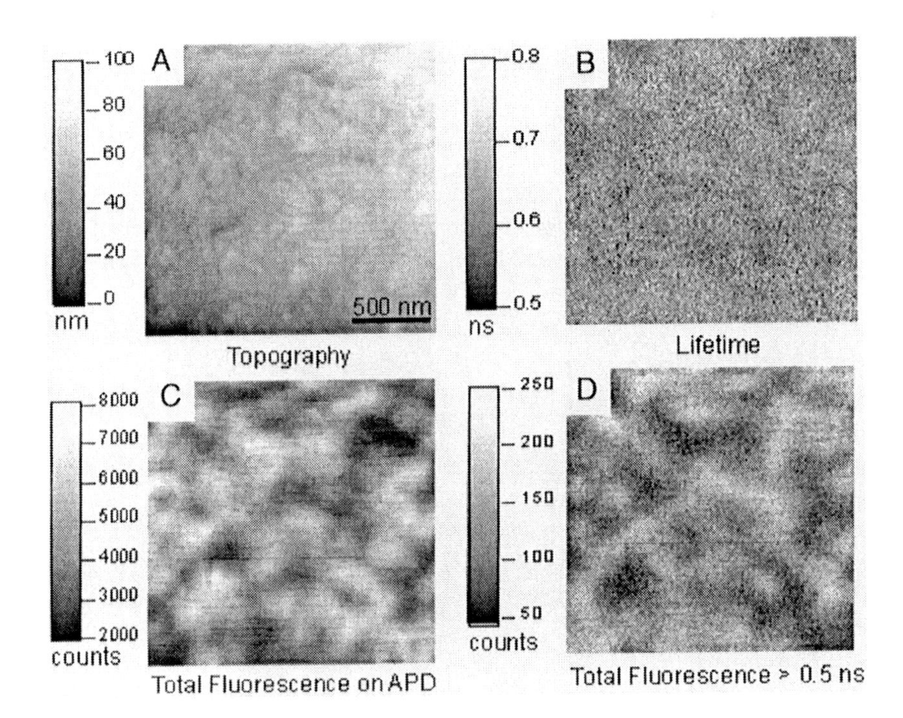

Figure 2: Time-resolved fluorescence NSOM images of annealed poly (9,9′–dihexylfluorene) thin film (5 x 5 μm). (A) Topography, (C) Total Fluorescence, (D) Total Fluorescence on MCP detector, which is orthogonal polarization to (C), and (B) Lifetime calculated using equation 1.

2D, should show different lifetimes in different polarizations. However, figure 2B is the lifetime image calculated using equation 1 and it shows no contrast.

This means that the emitting species are distributed evenly throughout the entire film and most importantly the 200 ps and 7 ns species emit with the same polarization. Previous models have predicted the long lifetime red emitting species result from excimer formation and emit with a polarization perpendicular to the polymer chain. The short wavelength lifetime 200 ps is polarized along the polymer backbone. Clearly the long time species, emit with this same polarization.

Metal Quenching

DPPE thin films have a singlet absorption peaked in the ultraviolet at 390 nm, while the emission is broad and structureless and peaked in the visible green near 540 nm.[2] Fluorescence lifetime measurements of pristine DPPE films exhibit nonexponential fluorescence decays with an average lifetime of approximately 2 ns.[16] Figure 3 shows a 2x2 μm NSOM image of a DPPE film deposited over a metal/glass interface. The topography (3A) of the polymer film is uniform with only a few minor clump-like defects on the sample surface that otherwise exhibits an rms roughness of only 7 nm. The metal boundary can be seen easily in the topographic image of the NSOM scan as a vertical cliff on the edge of a 40 nm thick plateau. The region of the sample possessing a metal film is also evident when examining the total fluorescence signal of the NSOM scan (figure 3). Since the polymer film is excited over the metal film and the fluorescence must subsequently travel through the metal layer before collection, the total fluorescence signal is significantly attenuated in metal coated regions. The resoultion, defined by the NSOM tip aperture, controls the sharpness of this boundary between light and dark regions. In this study it is approximately 100 nm.

The FLI-NSOM image in figure 3C is calculated using equation 1. In the fluorescence lifetime image, it is evident that their are two distinct regions with differing fluorescence lifetimes. Since lifetime decays from a single pixel consist of only a few hundred photons (Inset Figure 4), the signal to noise can be greatly increased by summing the counts from separate pixels in the same region of interest. 1000 pixels were averaged in both the long and short lived regions of the image to produce decays where accurate determinations of a $\langle\tau\rangle$ value for the polymer film in both regions could be made (Figure 4). The fluorescence decays are not single exponential due to the many conformations adopted by the large disorganized polymer molecules in the glassy spin cast films. The large number of random conformations provide a variety of differing environments and degrees of interpolymer interaction. Eventhough the decays are not single exponential, it is possible to compare the decays obtained from the two different

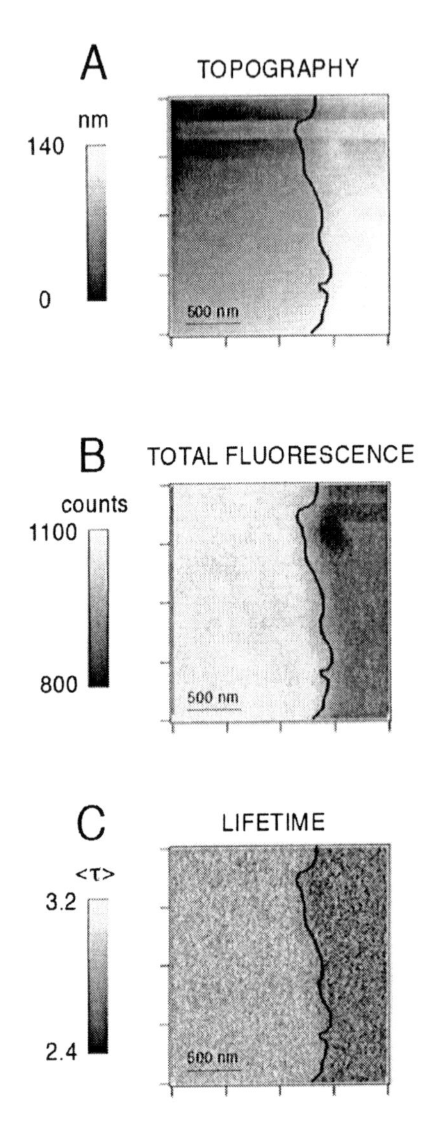

FIGURE 3. A single 2 x 2 um FLI-NSOM data scan provides A) topographic B) total fluorescence and C) lifetime images of a DPPE/gold interfacial region. The metal interface was traced in image A and superimposed on to both images B and C.

film regions. The decays from the two regions differ only beyond 1.63 ns. T=1.63 ns was subsequently used as the cuttoff time in the fluorescence lifetime image shown in Figure 3C to provide the greatest image contrast between the two regions of the sample.

Three main points should be stressed concerning the fluorescence lifetime image. The first, and most obvious, is that the contrast provided by FLI-NSOM is substantial, even for the small difference in $<\tau>$ calculated for the two regions of the polymer thin film. The subtlety of the lifetime quenching over the metal surface can be attributed to averaging of the lifetime information in the z-direction in the 40 nm thick film. Excited fluorophores may be quenched quite severely directly at the metal boundary but this information is averaged with lifetime information from other fluorophores in the cross-section of the film that

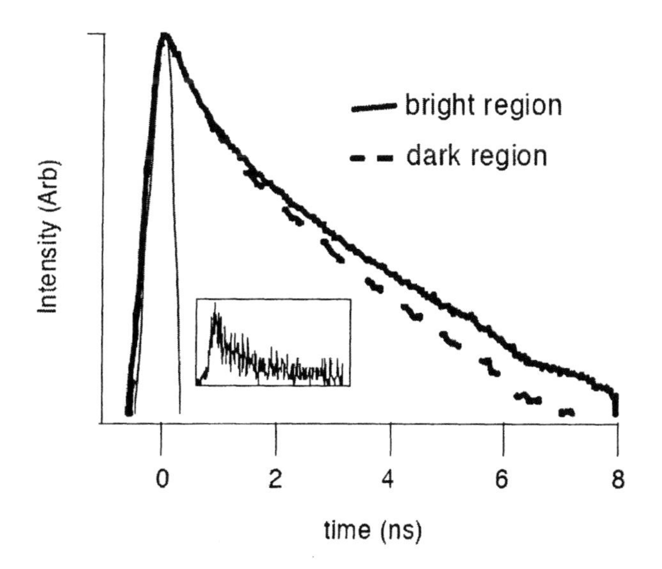

FIGURE 4. Log plots of fluorescence decays constructed by averaging 1000 pixels in the dark region (- - -) and the bright region (-----) of Figure 1C. The instrument response is also plotted. Inset: A fluorescence decay from a single image pixel.

are located farther from the metal surface and are less prone to nonradiative quenching interactions. The overall effect is a quenched decay that differs only slightly at times beyond 1.63 ns as compared with the unquenched system.

The second major aspect of the fluorescence lifetime image is the solid black line spanning the image from top to bottom. This line represents the superimposed boundary of the metal film in the plane of the sample scan and it can be drawn because of the direct correlation between surface and optical information during the NSOM scan. The boundary is determined by tracing the metal interface in the topographic image. The superior sensitivity and contrast of FLI-NSOM enables the detection of nonradiative fluorescence quenching beyond this boundary and into the bulk polymer region of the sample. The quenching effect, however, remains difficult to quantify as the quenching region is not of uniform penetration and the effect is small. The ability to detect this quenching effect but not to resolve its dimensions indicates that the quenching behavior beyond the metal interface is less than the resolution of our tip and can be estimated to be effective on a maximum distance between 20 and 50 nm, in good agreement with the data obtained concerning quenching over the metal surface itself in the 40 nm thick polymer film. The ability to detect fluorescence lifetime quenching at all on the small length scales involved is a testament to the sensitivity of FLI-NSOM as an analytical tool.

The third interesting characteristic of the fluorescence lifetime image concerns the chemical specificity of FLI-NSOM. When examining total fluorescence signals, scanning artifacts and variations in sample thickness often present problems concerning data interpretation. Clump-like areas of conjugated polymers often appear less intense in total fluorescence intensity images. However, as is seen in polyfluorene films the total intensity can be affected by many factors. Due to changes in total fluorescence, scanning artifacts often complicate data analysis along topological boundaries or interfaces. Small changes in the distance between the sample surface and the near-field aperture often lead to dramatic changes in overall fluorescence yields. These small changes in distance often occur when scanning over large objects in the tip path when shear force tracking varies slightly. When examining the images in Figure 3 it is easy to identify both of these types of phenomena. There is a slight glitch in the tip tracking in image 3A near the top of the scan and a cluster of clump-like structures that have low total fluorescence intensity are visible in image 3B. In image 3C, however, it is apparent that neither variation due to sample thickness nor to scanning artifacts have caused significant lifetime image contrast. Lifetime information provides for the clear identification of two distinct sample regions: unquenched polymer fluorescence over glass, and quenched polymer fluorescence over a gold surface.

Conclusion

The current study has demonstrated the viability of FLI-NSOM as a powerful analytical tool. Two main aspects of FLI-NSOM were highlighted in this work. The imaging technique provides high contrast information that allows for minor changes in fluorescence lifetime to visualized easily, and the NSOM lifetime images are less subject to artifact than the total fluorescence images. Using FLI- NSOM, lifetimes of both pristine and annealed film of PFH and metalized DPPE films have been measured. Studies indicate that lifetime is extremely uniform in both films despite small heterogeneities observed in the topography of the samples. The only lifetime contrast to be observed is the result of deliberate photobleaching of the film in the case of the polyfluorene or due to a metal layer in the case of the DPPE. The resulting lifetimes are much less subject to artifacts than the total fluorescence emission images. Although the method is sensitive to small chemical and environmental differences it is not, however, complicated by variations in sample thickness or other scanning artifacts. This stability allows for accurate interpretation of high resolution images without the pitfalls common to total fluorescence intensity measurements. While many factor can affect the total intentsity detected in a NSOM scan, there are only a few things that can alter the lifetime of an excited state. These include emission from distinct chemical species as is observed for the fluoronone defects in the PFH and energy transfer (quenching) as is seen in the DPPE samples.

Acknowledgements

The authors would like to thank Uwe Bunz for the DPPE material used in this study. This work supported by the National Science Foundation under Grant No. 9875315 and the by the Robert A. Welch Foundation grant F-1377. DAVB is a Research Corporation Cottrell Scholar and a Alfred P. Sloan Foundation Research Fellow.

References

(1) Heeger, A. J. Synth. Met. **2002**, 125, 23.
(2) Bunz, U. H. F.; Enkelmann, V.; Kloppenburg, L.; Jones, D.; Shimizu, K. D.; Claridge, J. B.; zur Loye, H.-C.; Lieser, G. Chem. Mater. **1999**, 11, 1416.
(3) Weder, C.; Sarwa, C.; Montali, A.; Bastiaansen, C.; Smith, P. Science **1998**, 279, 835.
(4) Grice, A. W.; Bradley, D. D. C.; Bernius, M. T.; Inbasekaran, M.; Wu, W. W.; Woo, E. P. Appl. Phys. Lett. **1998**, 73, 629.
(5) Paesler, M. A.; Moyer, P. J. Near-Field Optics: Theory, Instrumentation, and Applications; John Wiley and Sons: New York, 1996.

(6) Kwak, E.-S.; Kang, T. J.; Vanden Bout, D. A. Anal. Chem. **2001**, 73, 3257.
(7) Kwak, E.-S.; Vanden Bout, D. A. Analytica Chemica Acta **2003**, 496, 259.
(8) Chance, R. R.; A., P.; Silbey, R. Adv. Chem. Phys, **1978**, 37, 1.
(9) Teetsov, J.; Vanden Bout, D. A. Langmuir **2002**, 18, 897.
(10) Valaskovic, G. A.; Holton, M.; Morrison, G. H. Appl. Optics **1995**, 34, 1215.
(11) Karrai, K.; Grober, R. D. Ultramicroscopy **1995**, 61, 197.
(12) Lakowicz, J. R. Topics in Fluorescence Spectroscopy; Plenum press: New York, 1991; Vol. 1: Techniques.
(13) Teetsov, J.; Vanden Bout, D. A. J. Phys. Chem B **2000**, 104, 9378.
(14) Gaal, M.; List, E. J. W.; Scherf, U. Macromolecules **2003**, 36, 4236.
(15) Teetsov, J.; Fox, M. A. J. Mater. Chem. **1999**, 9, 2117.
(16) Imhof, J. M.; Bunz, U.; Vanden Bout, D. A. **2004**, manuscript in preparation.

Chapter 3

Exploring Dynamics in Photorefractive Polymer-Dispersed Liquid Crystals Using Near-Field Scanning Optical Microscopy

Jeffrey E. Hall and Daniel A. Higgins*

Department of Chemistry, Kansas State University, Manhattan, KS 66506

Dynamic near-field scanning optical microscopy (NSOM) imaging methods are employed to study liquid crystal dynamics in photorefractive polymer dispersed liquid crystal (PDLC) films. Micrometer-sized nematic droplets doped with photoexcitable electron donor and acceptor dyes and encapsulated within a thin polymer film are studied. Liquid crystal reorientation dynamics and ion migration dynamics are induced by applying a modulated electric field between the metalized NSOM probe and the electrically conductive, optically transparent substrate upon which the sample is supported. The induced dynamics are detected by monitoring the intensity of 633 nm light transmitted through the sample under crossed polarization conditions. Comparison of dynamics images taken before, during, and after photogeneration of ions using 488 nm light shows the influence of these ions on the liquid crystal dynamics. Computer simulations of the ion and liquid crystal dynamics are employed to better interpret the results. Taken together, these results indicate that differences observed between images recorded before (or after) and during ion generation result from changes in the liquid crystal reorientation dynamics deep within central droplet regions. In contrast, interfacial liquid crystal remains strongly aligned by the local space charge fields that develop. It is concluded that photorefractivity in PDLC films arises primarily from ion-induced relaxation of the liquid crystal in central droplet regions.

Introduction

Photorefractive materials are being developed for potential use in optical data storage *(1)* and processing *(2)*, security verification *(3)*, and medical imaging *(4)* technologies. A variety of photorefractive materials exist at present, ranging from inorganic crystals *(5)* to functionalized polymers *(6-8)*. Recently, significant effort has been devoted to the development of dye-doped liquid crystals as photorefractive systems *(9-12)*. Liquid crystals have a distinct advantage over solid-state photorefractives in that liquid crystals are strongly birefringent and can easily be reoriented by relatively small electric fields. The photorefractive effect in liquid crystal materials is known to arise from orientational relaxation of the liquid crystal brought about by the formation of space charge fields in the material. One disadvantage of bulk liquid crystal systems arises from the limited resolution that can be obtained *(13)*. This resolution limit stems from the increase in elastic energy that accompanies any distortion in the nematic phase alignment. One method by which this limitation can be overcome is by the use of polymer/liquid-crystal composites. Two classes of composite material have been employed; these are commonly known as polymer-supported liquid crystals (PSLCs) *(14)* and polymer-dispersed liquid crystals (PDLCs) *(13)*. In these materials, polymers are used to "divide" the liquid crystal phase into small, noninteracting domains *(13,15)*. While photorefractive PSLCs and PDLCs potentially provide greater resolution and have exhibited some of the largest photorefractive responses observed to date *(4)*, the addition of polymer leads to an increase in the field strength required to induce a response.

While the bulk properties of photorefractive polymer/liquid-crystal composites have been widely studied *(10,11,13,16)*, the mechanism by which photorefractivity occurs on a more localized level has only recently been investigated *(17)*. In our group, we have been studying the local dynamics associated with photorefractivity in PDLC systems using near-field scanning optical microscopy (NSOM) *(17,18)*. NSOM can be used to simultaneously obtain high-resolution topography, optical images and/or data on liquid crystal reorientation dynamics with sub-diffraction limited resolution *(19)*. Therefore, it represents a valuable means by which the mechanism of photorefractivity in micro/mesostructured composites can be explored.

In this paper, we summarize the results of both experimental and theoretical studies performed in our group on photorefractive polymer/liquid-crystal composites and present new results in support of our previous conclusions *(17)*. The specific samples studied are films formed from aqueous emulsions of a nematic liquid crystal mixture (E7) and poly(vinyl alcohol). Perylene and N,N'-di(n-octyl)-1,4,5,8-naphthalenediimide (NDI) are dissolved in the liquid crystal and serve as the photoexcitable electron donor and acceptor, respectively. This system was chosen because similar materials have been shown to exhibit strong photorefractivity *(10)*. In the discussion below, we start by presenting results from bulk spectroscopic studies that show these materials possess the photophysical attributes necessary for photorefractivity. Next, we report results from our measurements of the photorefractive gain coefficient for PSLC materials. Data from extensive NSOM studies of similar PDLCs are then presented and discussed in detail. These results provide spatially-resolved information on the influence of photogenerated ions on the field-induced reorientation process in the liquid crystal droplets. Finally, the results of computer simulations are used to assist in the interpretation of the NSOM data.

Experimental

Bulk Spectroscopy. Fluorescence excitation and emission spectra for perylene (Aldrich) and NDI (synthesized as described in the literature *(20)*) were obtained using a conventional fluorimeter. Chloroform solutions of perylene and NDI at 20 μM concentrations were employed. Investigations of perylene fluorescence quenching by NDI were performed using a series of 20 μM perylene solutions with varying concentrations of NDI (1-10 mM). In these studies, 457 nm light was used to excite perylene fluorescence.

Two-Beam Coupling. Quantitative measurements of the photorefractive gain coefficient in PSLCs were obtained from asymmetric beam coupling experiments. A detailed explanation of the instrumentation and methods used can be found in our original publication *(21)*. Briefly, photorefractive PSLC cells were prepared from dye-doped liquid crystal emulsions formed in aqueous poly(vinyl alcohol) (PVA) solutions. These samples were prepared as follows. First, the liquid crystal (E7, Merck) was doped with perylene (electron donor) and NDI (electron acceptor) to concentrations of 2.1 mM and 6.8 mM, respectively. The dye-doped liquid crystal was then emulsified in a 2% (by weight) PVA solution to yield a 9:1 liquid crystal/polymer ratio (by weight). Free ions had been removed from the PVA solution prior to use by dialysis against several portions of ultrapure water (18 MΩ•cm). Sample cells were fabricated by placing a 20 μm thick Mylar spacer onto an indium-tin-oxide (ITO) coated glass substrate. A 1 cm^2 hole was cut in the center of the spacer to form the working cell volume. A drop of the polymer/liquid-crystal emulsion was deposited onto the substrate and the water allowed to evaporate. A second ITO substrate was then placed on top of the spacer. Finally, electrical connections were made to each ITO slide using fine transform wire, silver paint, and super glue.

For two-beam coupling measurements, the sample cell was mounted vertically at the intersection of two p-polarized beams (488 nm) from an argon-ion laser. Diffraction of the transmitted beams by the index grating formed in the sample was measured using photodiode detectors. Data was recorded on a computer as a function of applied field, as one of the two incident beams was chopped. A waveform generator provided the applied electric field, which consisted of a modulated (5 kHz) field superimposed on a dc field. Lock-in detection of the chopped, diffracted beams allowed for direct measurement of the asymmetric beam coupling efficiency *(21)*.

NSOM Imaging. The samples prepared for use in NSOM experiments incorporated significantly more polymer than those used to observe bulk photorefractivity and are appropriately characterized as PDLC films. Here, dye-doped liquid crystal was emulsified with a 3% (by weight) purified PVA solution to form a 3% (by weight) liquid-crystal-in-polymer emulsion. This emulsion was then spin cast onto an ITO substrate and dried. Electrical connection to the substrate was made as described above.

The influence of photogenerated ions on the electric-field-induced reorientation dynamics in single liquid crystal droplets was studied using a modified TM-Microscopes Aurora NSOM (Veeco Metrology). Details of these experiments may be found elsewhere *(17)*. but are briefly outlined below. NSOM fiber probes were made in-house for these studies *(22)*. Static and dynamic optical images of the droplets were recorded along with topographic images. Tuning-fork-detected shear-force feedback methods were employed to maintain the tip-sample separation at ≈5 nm during imaging *(23)*.

For the recording of dynamics images, a sinusoidally modulated (200 Hz) electric field was applied between the NSOM probe and sample substrate. Electrical connection was made to the end of the metal-coated probe using fine transformer wire, silver paint and super glue. Field-dependent changes in the liquid crystal orientation state were detected by monitoring the transmission (under crossed polarization conditions) of 633 nm light exiting the probe and passing through the sample. Transmitted 633 nm light was detected using a PMT. Lock-in detection of the modulated PMT signal provided amplitude and phase data related to the liquid crystal reorientation dynamics. These signals were fed directly into the microscope control electronics to record dynamics images. Such images were acquired before, during and after photogeneration of ions. Ion generation was accomplished using 488 nm light (with 50-100 pW exiting the probe) to excite the dissolved perylene. For images recorded during ion generation, 633 nm light and 488 nm light were simultaneously coupled into the NSOM probe fiber. Residual 488 nm light was blocked in the collection path of the microscope using a holographic notch filter and a dichroic mirror. Images recorded before and after ion generation employed 633 nm light only.

Results and Discussion

Bulk Spectroscopy. Excitation and emission spectra for perylene and NDI dissolved in chloroform are shown in Figures 1A,B. These spectra are characteristic of the monomeric dyes and show no evidence for aggregation at the concentrations employed. Indeed, absorption spectra recorded for NDI in chloroform at concentrations as high as 1.2 mM showed no evidence for

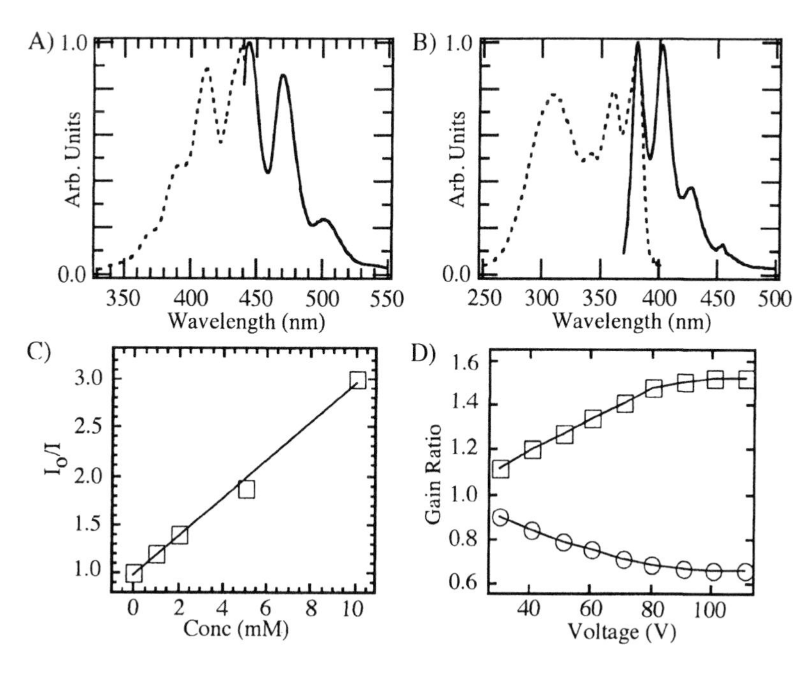

Figure 1. A) Perylene excitation (dashed) and emission (solid) spectra. B) NDI excitation (dashed) and emission (solid) spectra. C) Stern-Volmer plot of perylene fluorescence quenching vs. NDI concentration. D) Photorefractive gain ratio from asymmetric two-beam coupling experiment.

aggregation. Hence, it may be concluded that the effects described below result from interactions between perylene and NDI monomers.

The results of perylene fluorescence quenching studies performed as a function of NDI concentration are shown in Figure 1C. As depicted, perylene fluorescence is strongly quenched by NDI. Fluorescence quenching is attributed here to transfer of an electron from perylene to NDI. Perylene quenching is not expected to occur via energy-transfer to NDI because perylene emits well to the red (440-525 nm) of the NDI excitation spectrum (250-390 nm).

The rate of electron transfer quenching between perylene and NDI was determined from the data in Figure 1C. The slope of the line fit to this data, using the Stern-Volmer Equation, gives a quenching rate constant of approximately 10^{10} $M^{-1}s^{-1}$. Such a large value is consistent with quenching at the diffusion-limited donor-acceptor collision rate.

Two-Beam Coupling. Verification of photorefractivity in the polymer/liquid-crystal composites was obtained from asymmetric two-beam coupling experiments *(21)*. In these studies, two mutually coherent 488 nm laser beams were overlapped in the sample cell, creating an optical intensity grating. Electron transfer between the perylene and NDI in regions of constructive interference produces photogenerated charges in these same regions. These charges migrate by both diffusion and field-induced drift mechanisms, the latter occurring with the aid of the applied (dc) electric field. Internal space-charge fields are formed as a result, modifying the total electric field at each point in the sample and causing the liquid crystal to reorient. A refractive index grating that is spatially displaced from the original intensity grating is formed. This index grating causes light from the two incident lasers beams to be preferentially diffracted in one direction.

The results of two-beam coupling measurements on perylene and NDI-doped PSLCs are shown in Figure 1D. These data show a field-dependent increase in one of the two beams (squares) and a concomitant decrease in the other (circles), as expected. With certain assumptions *(21)*, this data can be used to determine the photorefractive gain coefficient for the sample *(10)*. Here, it is determined to be 808 cm^{-1}, close to that determined previously for similar materials *(10)*. This result demonstrates the materials chosen for study herein exhibit large photorefractive responses compared to many solid state organic and inorganic materials *(4)*.

NSOM Imaging. NSOM experiments were performed to obtain a better understanding of the local ion migration and liquid crystal reorientation dynamics associated with photorefractivity in dye-doped polymer-dispersed liquid crystal (PDLC) films *(17)*.

In dynamic NSOM imaging experiments, a sinusoidally modulated electric field is applied across the sample, between the NSOM probe and ITO-coated substrate. This field produces a time-dependent modulation of the liquid crystal orientation in droplets positioned below the probe. In addition, this same field causes ions dissolved in the liquid crystal to migrate to/from the polymer/liquid-crystal interfaces. Charged interfacial double layers are formed at these interfaces under the influence of the applied field. The buildup and decay of these double layers modifies the local electric field within the droplet in time, causing an additional change in the liquid crystal orientation state.

Changes in liquid crystal orientation are detected optically, by monitoring the field-modulated birefringence of the liquid crystal. This is accomplished experimentally by measuring the intensity of 633 nm light transmitted through the sample under cross-polarization conditions. A modulated optical signal results, for which the amplitude and phase characteristics are strongly dependent on the local concentration of ions and the viscoelastic properties of the droplet *(19)*. A lock-in detector is used to obtained amplitude and phase signals, and these are in turn used to record dynamics images. These images provide a direct means of observing spatial variations in the liquid crystal reorientation dynamics in the presence or absence of permanent and/or photogenerated ions dissolved in the PDLC droplets *(19,24,25)*.

Dynamic NSOM amplitude and phase images of a single droplet in a photorefractive PDLC film are shown in Figure 2. Again, experiments were conducted by imaging the droplet before (633 nm only), during (488 nm and 633 nm), and after (633 nm only) photogeneration of ions. The phase images provide the simplest means for observing the overall (i.e. average) effects of photogenerated ions on the liquid crystal dynamics in single droplets. Likewise, the amplitude images provide the best means for observing spatial variations in the dynamics caused by changes in the ion concentration during the recording of sequential images of a droplet.

The phase data presented in Figure 2 clearly shows the influence of photogenerated ions on the liquid crystal dynamics. The mean phase angle over the entire droplet in Figure 2A is 175.8±0.9° (488 nm off), prior to ion generation. The error bars given depict the 95% confidence interval. During ion generation, the phase angle increases to 179±1° (488 nm on), as shown in Figure 2D. Following ion generation, the phase angle returns to 174.0±0.8° (488 nm off), as in Figure 2G. It should be noted that undoped PDLC materials do not show such significant changes in otherwise identical experiments. The observed increase in phase angle indicates faster liquid crystal dynamics occur in the presence of ions, as expected. Faster dynamics may result from either enhanced field-driven reorientation of the liquid crystal, or an enhanced rate of decay of the local electric field, and hence, enhanced relaxation of the liquid crystal towards it zero-field alignment. At present, computer simulations are needed to distinguish which of these effects is most important. The results of such simulations are discussed in detail, below. The return of the phase angle to near its original value after ion generation indicates that most of the photogenerated ions recombine by back electron transfer or are removed by some other process on the time scale of a single image (i.e. ten minutes).

While the phase images provide clear evidence for the influence of photogenerated ions on the average droplet dynamics, only subtle spatial variations are observed in these images. The relatively weak spatial contrast in the phase data may indeed be reflective of relatively uniform ion-modulated reorientation dynamics within the droplet. However, since the amplitude images suggest otherwise (see below), it may simply indicate reduced signal-to-noise in the phase data, or a reduced sensitivity to small changes in the dynamics.

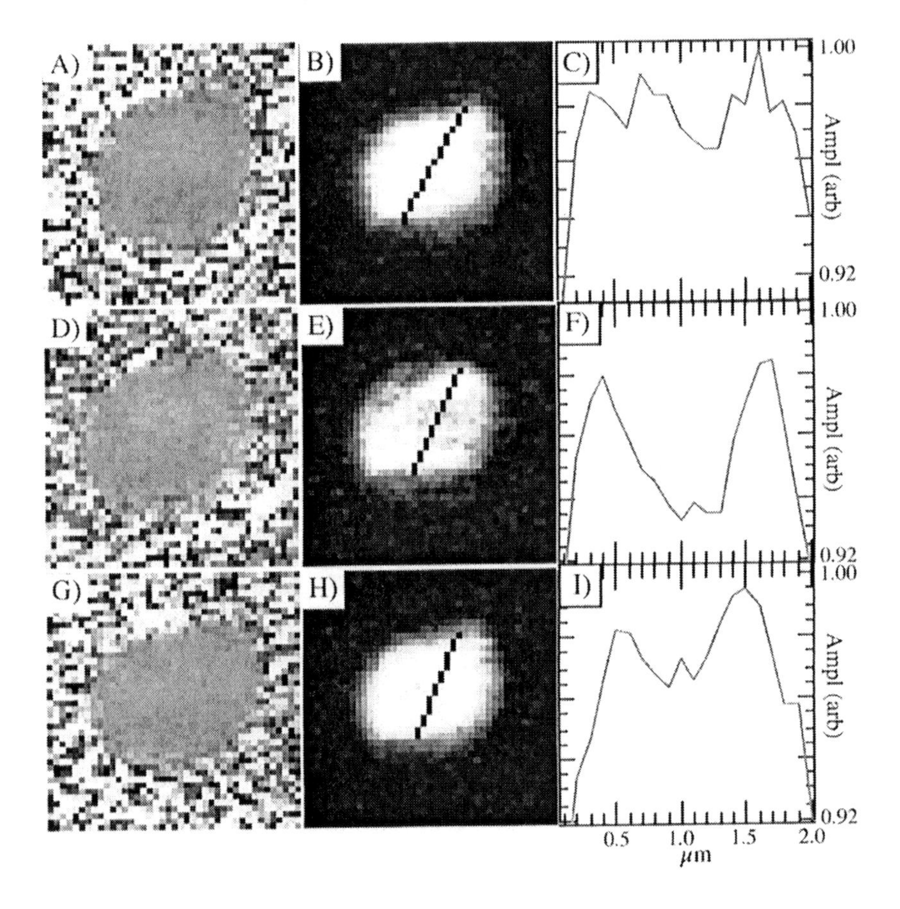

Figure 2. A) Phase image and B) amplitude image of an ≈2.8 μm diameter photorefractive droplet imaged using 633 nm light. C) Line scan from image B). D) Phase image and E) amplitude image of same droplet, imaged during photogeneration of ions using 488 nm light. F) Line scan from image E). G) Phase image and H) amplitude image after photogeneration of ions. I) Line scan from image H).

The amplitude images shown in Figures 2B,E,H clearly depict spatial variations in the ion-modulated liquid crystal reorientation dynamics. Figures 2C,F and I depict line scans taken from these images and illustrate some of the salient changes in the reorientation dynamics more clearly. Most importantly, during ion generation, the amplitude signal in the center of the droplet decreases relative to that observed in outer droplet regions. Such a decrease in the amplitude signal is indeed expected to occur with an increase in ion concentration (see below). Here, it is proposed that the observed spatial variations in the amplitude signal caused by the photogenerated ions arise from the formation of highly nonuniform charge distributions (laterally) at the upper and lower droplet surfaces, nearest the electrodes. As these interfaces are not planar, but rather more closely approximate the internal surfaces of a sphere, it is believed the ions tend to migrate to the uppermost and lowermost regions of the droplet, leading to formation of a nonuniform space charge field within the droplet. The largest space charge fields occur along the central axis of the droplet and decay rapidly as a function of radial distance towards the outer (circumferential) droplet regions. Hence, the applied field and liquid crystal dynamics are perturbed to a smaller extent in outer droplet regions. The effects of such variations are amplified by the dependence of liquid crystal reorientation on the square of the local field.

A number of other possible explanations exist as well; these have been discussed previously *(17)*. They include the possibility that the relative importance of the local field and the droplet's viscoelastic properties may vary spatially as well. The change in amplitude over the center of the droplet is consistent with the greater importance of the local field dynamics in this region, while the elastic properties of the droplet are clearly more important in circumferential droplet regions *(26)*. Another possibility is that more ions are generated with the probe positioned over the center of the droplet, due simply to the greater optical thickness in this region.

Simulations. As described previously *(17)*, finite difference time domain methods can be used to simultaneously model the ion dynamics, local space charge field dynamics, liquid crystal reorientation dynamics, and the observed optical signal in photorefractive PDLC droplets. Such simulations were performed for a simplified one-dimensional model of a droplet. The dynamics were simulated along a line running parallel to the optical path through the droplet center (i.e. normal to the film plane). A constant number of permanent ions were incorporated in the droplet, an approximation meant to simulate steady-state conditions during ion generation. The droplet was 1.0 μm in diameter and was surrounded by a 20 nm thick dielectric "shell", representing the polymer film covering real droplets. Known optical and dielectric parameters associated with the polymer and liquid crystal were employed *(17)*.

Figure 3 shows some results from these simulations for droplets incorporating ions at 1 μM and 30 μM concentrations (defined prior to application of the field). The applied field in these simulations was modulated at 200 Hz, as in the NSOM experiments. The simulations were run for 14 field cycles to achieve a steady-state response. On the simulation time scale, the applied electric field remains relatively uniform throughout the droplet, at low ion concentrations (1 μM). In contrast, the droplet containing 30 μM ions shows

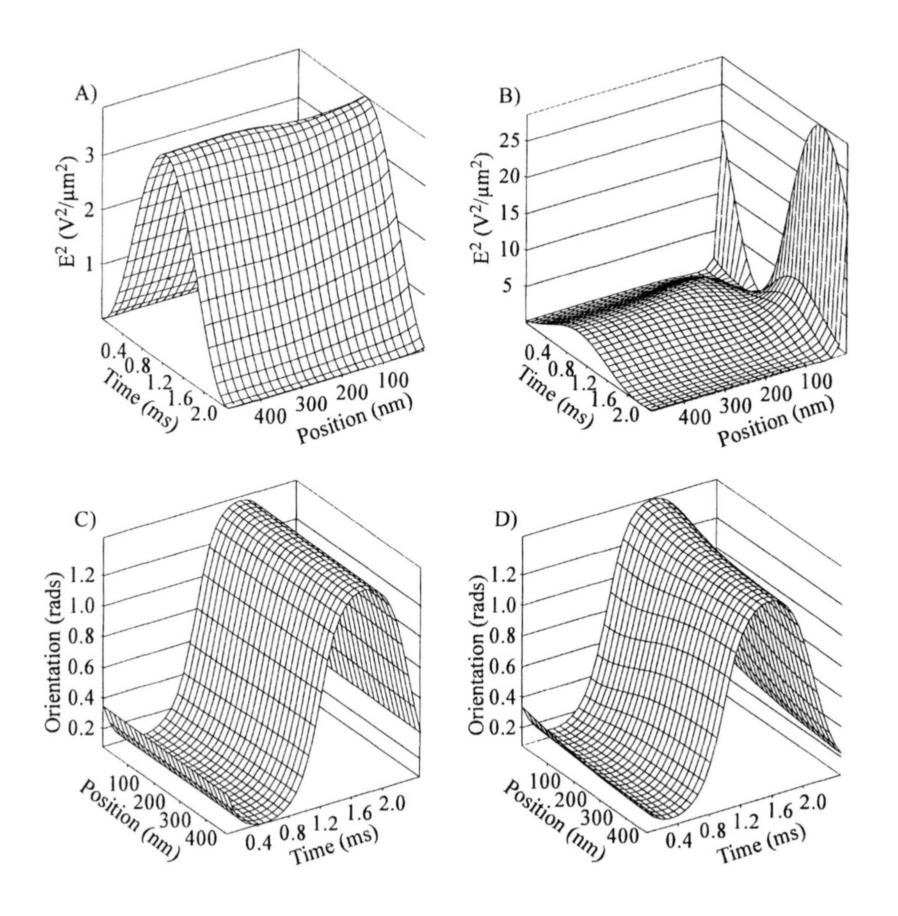

Figure 3. Squared electric field in a droplet with A) 1 μM ions and B) 30 μM ions, as a function of time and position within the droplet. Ion migration dynamics are induced by an externally applied sinusoidally modulated electric field. C) and D) Liquid crystal orientation angle as a function of time and position within the droplet resulting from the electric fields shown in A) and B). Position within the droplet is given as a function of distance from the polymer/liquid-crystal interface at 0.0 nm. These data were obtained after 13 field cycles in the simulations.

dramatic spatial and temporal variations in the internal electric field. The field rises and falls in the droplet center but never gets very large. In contrast, the field at the polymer/liquid-crystal interface becomes very large as the field in the center of the droplet decreases towards zero. This result indicates liquid crystal reorientation is alternately driven by I) a weak field in the center of the droplet and II) a strong field that builds later, in the interfacial regions. Due to the collective response of the liquid crystalline phase, the actual reorientation dynamics are much more homogeneous, even in droplets with high ion concentrations (see Figures 3C,D). Nevertheless, due to the time-dependent decay of the field in the center of the droplet, orientational relaxation of the liquid crystal does indeed occur in the droplet center (see Figure 3D). The liquid crystal near the polymer/liquid-crystal interface, however, remains in a strongly field-aligned state. Therefore, it may be concluded that the changes observed in the dynamics images during ion generation (see above) arise from ion-dependent changes in the liquid crystal reorientation dynamics predominantly "deep" in the central regions of the droplet. It may also be concluded that the orientational relaxation associated with the photorefractive effect in bulk PDLC systems also occurs in these same droplet regions, with interfacial liquid crystal dynamics of less importance.

Finally, as a means to demonstrate that the simulations and NSOM data are reflective of the same dynamical phenomena, the optical signals obtained in NSOM experiments were also modeled *(17)*. The results are shown in Figure 4. It is clear from these data that the modulated optical signal is predicted to shift

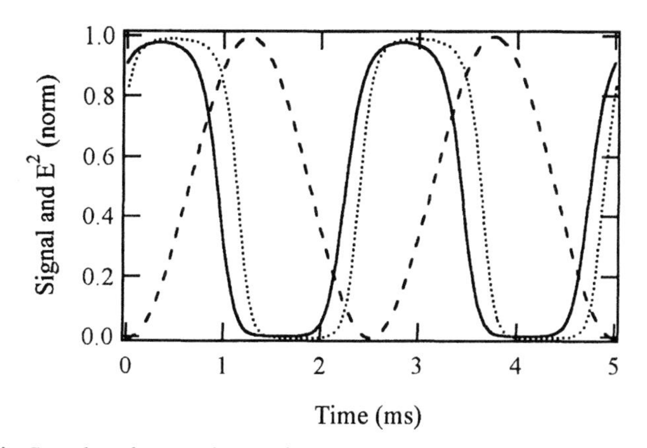

Figure 4. *Simulated optical signal and squared applied field (- - - -) for a 1 μm thick liquid crystal droplet incorporating 1 μM ions (······) and 30 μM ions (——). The optical signal is near 1.0 in the relaxed liquid crystal state and 0.0 in the field-aligned state.*

earlier in time (i.e. a positive phase shift occurs) with increasing ion concentration, exactly as observed in the NSOM data. The simulated signals also exhibit a decrease in amplitude with increasing ion concentration, as observed experimentally. Most importantly, these results indicate that the experimentally observed phase shift arises primarily from an increase in the rate of field-induced liquid crystal reorientation. The enhanced reorientation rate is caused by the large fields that develop in the interfacial regions at high ion concentrations. The increased decay rate of the electric field in the droplet center also causes the liquid crystal to relax towards its zero-field orientation earlier in time, although the contribution of this effect to the phase shift is smaller (i.e. compare the relative time shifts in the rising and falling edges of the simulated signals in Figure 4).

Conclusions

This article has presented a review of our recent bulk spectroscopy and NSOM studies of photorefractive PDLC materials. New NSOM data and simulation results were also given. These support our previous conclusions on the origins of local phenomena associated with photorefractvity in PDLC films *(17)*. Bulk spectroscopic studies showed that photoinduced electron transfer occurs between perylene and NDI at the bimolecular collision rate. Asymmetric beam coupling experiments showed a strong photorefractive effect in PSLC cells fabricated from perylene and NDI-doped polymer/liquid-crystal composites. Dynamic NSOM experiments provided important new information on the ion-dependent, field-induced reorientation dynamics within single liquid crystal droplets in similar PDLC films. Finite difference time domain simulations of the dynamics within the droplets provided information necessary to draw clear conclusions from the NSOM data. The experimental data and simulation results indicate that changes in the liquid crystal dynamics observed in the presence of ions occur primarily deep within central droplet regions, while the interfacial liquid crystal remains in a strongly field-aligned state. Taken together, the results of the simulations and the NSOM data provide a valuable new means for studying the local dynamics associated with photorefractivity in PDLC based materials.

Acknowledgements

The authors thank the NSF (CHE-0092225) for funding this work.

References

1. Cipparrone, G.; Mazzulla, A.; Nicoletta, F.; Lucchetti, L.; Simoni, F. *Optics Communic.* **1998**, 297.
2. Ono, H.; Kawatsuki, N. *SPIE Proc.* **1998**, *3475*, 122.
3. Kippelen, B.; Volodin, B.; Meerholz, K.; Javidi, B.; Peyghambarian, N. *Nature* **1996**, *383*, 58.
4. Zilker, S. *Chem. Phys. Chem.* **2000**, *1*, 72.
5. Ashkin, A.; Boyd, G. D.; Dziedzic, J. M.; Smith, R. G.; Ballman, A. A.; Levinstein, J. J.; Nassau, K. *Appl. Phys. Lett.* **1966**, *9*, 72.
6. Moerner, W. E.; Silence, S. *Chem. Rev.* **1994**, *94*, 127.
7. Van Steenwinckel, D.; Hendrickx, E.; Persoons, A. *J. Chem. Phys.* **2001**, *114*, 9557.
8. Yu, L.; Chan, W. K.; Peng, Z.; Gharavi, A. *Acc. Chem. Res.* **1996**, *29*, 13.
9. Khoo, I. C.; Li, H.; Liang, Y. *Optics Lett.* **1994**, *19*, 1723.
10. Wiederrecht, G.; Yoon, B.; Wasielewski, M. *Science* **1995**, *270*, 1794.
11. Cipparrone, G.; Mazzulla, A.; Simoni, F. *Mol. Cryst. Liq. Cryst.* **1997**, *299*, 329.
12. Ono, H.; Kawatsuki, N. *Rec. Res. Dev. Appl. Phys.* **1998**, 47.
13. Golemme, A.; Kippelen, B.; Peyghambarian, N. *Appl. Phys. Lett.* **1998**, *73*, 2408.
14. Wiederrecht, G.; Wasielewski, M. *J. Am. Chem. Soc.* **1998**, *120*, 3231.
15. Wiederrecht, G.; Wasielewski, M. *J. Nonlin. Opt. Phys. Mater.* **1999**, *8*, 107.
16. Ono, H.; Kawatsuki, N. *Appl. Phys. Lett.* **1997**.
17. Hall, J. E.; Higgins, D. A. *J. Phys. Chem. B* **2003**, *107*, 14211.
18. Dunn, R. C. *Chem. Rev.* **1999**, *99*, 2891.
19. Mei, E.; Higgins, D. A. *J. Chem. Phys.* **2000**, *112*, 7839.
20. Wiederrecht, G.; Svec, W.; Niemczyk, M.; Wasielewski, M. *J. Phys. Chem.* **1995**, *99*, 8918.
21. Hall, J. E.; Higgins, D. A. *Rev. Sci. Instrum.* **2002**, *73*, 2103.
22. Hall, J. E.; Higgins, D. A. *Poly. Mater. Sci. Eng.* **2003**, *88*, 186.
23. Karrai, K.; Grober, R. *Appl. Phys. Lett.* **1995**, *66*, 1842.
24. Higgins, D. A.; Liao, X.; Hall, J. E.; Mei, E. *J. Phys. Chem. B.* **2001**, *105*, 5874.
25. Mei, E.; Higgins, D. A. *J. Phys. Chem.* **1998**, *102*, 7558.
26. Higgins, D. A.; Luther, B. J. *J. Chem. Phys.* **2003**, *119*, 3935.

Chapter 4

Chemical Imaging of Heterogeneous Polymeric Materials with Near-Field IR Mircroscopy

Chris A. Michaels[1], D. Bruce Chase[2], and Stephan J. Stranick[1]

[1]Chemical Science and Technology Laboratories, National Institute of Science and Technology,, 100 Bureau Drive, Gaithersburg, MD 20899
[2]DuPont Central Research and Development, Wilmington, DE 19880

The development of techniques to probe spatial variations in chemical composition on the nanoscale continues to be an important area of research in the characterization of polymeric materials. Recent efforts have focused on the development and characterization of near-field microscopy in the mid-infrared spectral region. This technique involves coupling the high spatial resolution of near-field scanning optical microscopy with the chemical specificity of infrared absorption spectroscopy. This chapter describes the application of this technique to the chemical imaging of a polymer composite consisting of polystyrene particles dispersed in a poly (methyl methacrylate) thin film. Particular focus is placed on utilizing the measured spectral contrast to identify the nature and magnitude of the various sources of image contrast, including near-field coupling, topography induced optical contrast, scattering, and absorbance.

Introduction

The ability to routinely measure local variations in chemical composition on length scales finer than those accessible to conventional optical microscopy would be of significant utility in the characterization of complex polymeric materials such as blends and composites. Sample analyses of this type would be particularly useful in establishing correlations between macroscopic performance properties (*e.g.* mechanical and chemical stability, biocompatibility) with material microstructure, a key ingredient in the rational design of high performance materials. One strategy for realizing this goal involves the integration of vibrational spectroscopy into a near-field scanning optical microscope (NSOM). This combination of the sub-diffraction spatial resolution attainable in the near-field with the high chemical specificity of vibrational spectroscopy promises a powerful analytical technique that overcomes critical measurement limitations of both far-field vibrational microscopes (low spatial resolution) and scanned probe microscopes (lack of chemical specificity). Both Raman (1-8) and IR (9-18) spectroscopy have been exploited in near-field measurements; the optimum choice depending generally on the magnitude of the cross sections for the relevant optical transitions. The large body of literature on far-field IR microscopy of polymeric systems (19) suggests that these materials would be ideally suited for study with near-field IR microscopy, wherein spatial variations in material composition would result in image contrast due to the corresponding variations in infrared absorbance. The realization of infrared absorption spectroscopy as a source of near-field optical contrast would enable the mapping of chemical functional groups unique to individual polymers as a function of position with sub-diffraction spatial resolution (17).

In conventional optical microscopy, the spatial resolution is set by the diffractive properties of light, limiting the attainable resolution to approximately $\lambda/2$ (λ = wavelength of the light) (20). In 1928, Synge recognized that this resolution limit could be circumvented in what is now called a near-field microscope (21). One common implementation of near-field microscopy involves illumination of a sample through an aperture with a diameter significantly smaller than λ, while maintaining the sample-to-aperture separation at a distance much smaller than $\lambda/2$ (22). In this arrangement, an electromagnetic field confined to dimensions defined roughly by the aperture diameter interacts with the specimen, thus enabling the study of many types of photon-matter interactions (e.g. scattering (23), fluorescence (24), absorbance (17), and harmonic generation (25)) with spatial resolution unattainable in conventional far-field microscopy. In this contribution, the use of infrared near-field microscopy in the study of the microstructure of a polymer composite, consisting of polystyrene microspheres distributed in a thin layer of poly (methyl methacrylate) is described.

Although multiple reports of infrared NSOM have appeared in the literature (9-18), fundamental metrology questions about this technique persist. In particular, a general description of the nature and relative magnitude of the operative sources of contrast in the imaging of samples expected to absorb significant quantities of the incident radiation has yet to appear. The respective roles of the real and imaginary parts of the refractive index and their impact both on the measured spectra and attempts to extract chemical information from the images is not yet fully understood. This is particularly important in the vicinity of strong absorbance resonances where the real part of the refractive index can change dramatically over narrow frequency regions, as described by the Kramers-Kronig relations. This might lead to band shifting effects, raising the question of what the expected correlation is between far and near-field infrared spectra, with obvious ramifications for the assignment of near-field spectra. Additionally, the presence of multiple, spatially correlated sources of image contrast (e.g. topography and absorbance) can greatly complicate image interpretation and the development of robust methods for unraveling such multiple contrast mechanisms is only at a nascent stage (17). While comprehensive answers to these questions are clearly beyond the scope of this chapter, the near-field images and spectra of a polymer composite reported here will demonstrate the high spatial resolution, chemical imaging power of this technique, while simultaneously illuminating some of the aforementioned difficulties surrounding detailed image interpretation, perhaps stimulating further experimental and theoretical efforts along these lines.

Experimental

There are a number of technical challenges to the implementation of IR NSOM that are unique to the mid-IR spectral region. Particularly, near-field probes that transmit infrared radiation are required, as is a bright, tunable infrared radiation source. We have developed a novel bench-top infrared NSOM instrument that incorporates tapered optical fiber probes and a femtosecond laser as the infrared source. This instrument has been described in detail elsewhere (15,23) but a brief synopsis of the key features follows. The broadband infrared light source is based on a commercial system that includes a Ti:sapphire oscillator and 250 kHz regenerative amplifier that pumps a BBO optical parametric amplifier, yielding a signal beam (1.1 μm - 1.6 μm) with an 85 fs pulsewidth (as measured by background free autocorrelation) and an idler beam (1.6 μm - 2.5 μm). Mid-infrared light is generated by difference frequency mixing the signal and idler beams in a 1-mm thick $AgGaS_2$ crystal, cut for type-I mixing. This difference frequency generation scheme has a demonstrated tuning range of 2.5 μm to 12 μm. Average powers at 3.4 μm (the nominal operating wavelength for the work described here) range from 8 mW to 12 mW with an optimized spectral FWHM of approximately 150 cm^{-1}. The small fraction of light transmitted through the sub-wavelength aperture (typically 1 to 5 x 10^{-5}) is collected by a 0.5 NA, CaF_2 lens located behind the sample. Following collection, the sample and reference beams are steered and focused onto the entrance slit of a 0.32 m monochromator incorporating a 75 grooves/mm diffraction grating blazed at 4.5 μm. A 128x128 pixel InSb array detector is positioned at the exit focal plane of the monochromator to detect the horizontally dispersed reference and sample spectra, slightly offset from each other in the vertical direction. The full bandwidth of the laser light is collected in parallel with the focal plane array, eliminating the need to scan a grating or the laser wavelength.

The sample is scanned over the near-field probe with a Besocke style scanner in which a solid-state, piezoelectric detection method is used to measure the shear forces applied to the probe by the sample. In constant-gap mode (CGM) scanning, the probe-sample separation is kept constant using feedback regulation of the shear force signal, yielding a topographic map of the sample. In constant-height mode (CHM) scanning, the vertical position of the sample is set such that the probe is approximately 100 nm above the highest sample protrusion within the scan area; the horizontal scanning of the sample then proceeds with no active probe-sample distance regulation. Spectral image collection involves the raster scanning of the sample over the probe, coupled with parallel acquisition of infrared spectra from the focal plane array, synchronized to the probe position. Typical individual spectrum integration times range from 10 msec - 50 msec, and each spatial pixel involves an average of 20 spectra, thus yielding typical data acquisition times of 1 sec per point. The resultant three dimensional data set includes two spatial dimensions and a spectral dimension.

This illumination mode microscope is based on a near-field probe fabricated by tapering and metal coating an IR transmitting, single mode, fluoride glass optical fiber. The etching methodology is similar to that used in the fabrication of NSOM tips from glass and chalcogenide fibers (26,27). The stripped fibers are inserted several mm below the interface between the etchant solution (0.06 M HCl and 0.04 M $ZrOCl_2$) and a 2,2,4 trimethylpentane organic overlayer for 20 minutes. The apertures are formed by sequential vapor deposition of two 80 nm -100 nm thick Al or Ag films on rotating fiber tapers held at an angle of approximately 15° with respect to the deposition source. The imaging characteristics of each probe are determined through evaluation of the spatial resolution and image contrast present in near-field transmission images of a micropatterned gold test sample, as described previously (15).

The polymer composite sample was fabricated by spin casting a suspension of 1.1 μm diameter polystyrene (PS) microspheres in a solution of poly (methyl methacrylate) (PMMA) in toluene onto a microscope cover slip. The spin conditions were optimized to produce a polymer layer with a nominal thickness of 1 μm. Fourier transform transmission infrared spectra of pure PS and pure PMMA films were recorded on an infrared spectrometer (Nicolet Nexus 870) equipped with a HgCdTe detector (28).

Results and Discussion

Figure 1 shows infrared absorbance spectra of PS (dashed line) and PMMA (solid line) in the C-H stretching region along with the broadband illumination spectrum of the radiation (filled grey) used to acquire the images reported here. The magnitude of the absorbance spectra and the laser spectrum have been scaled to facilitate visual comparison. Although the spectral breadth of the laser source does not approach that of the blackbody sources typically employed by IR spectroscopists, it is sufficient to encompass multiple bands and, as is clear from Figure 1, it is broad enough to allow the differentiation of two polymeric materials. Images acquired over this spectral region include those at frequencies of a unique PS resonance (3060 cm^{-1}), resonant frequencies where both materials might be expected to absorb (2990 cm^{-1}, 3020 cm^{-1}) and a frequency where neither material has a significant absorbance (3120 cm^{-1}).

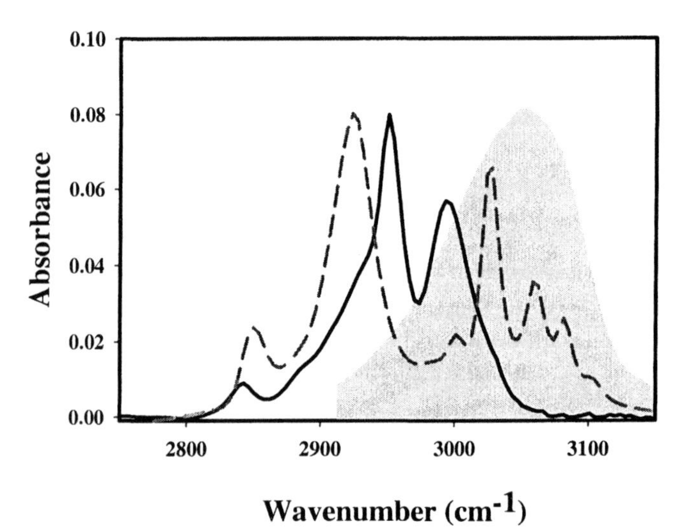

Figure 1. Scaled IR absorbance spectra of PS (dashed line) and PMMA (solid line) in the C-H stretch region along with with spectrum of imaging radiation.

Figure 2(a) is an 8 μm x 8 μm CGM topographic image showing several PS particles protruding from the film, where the full range of the linear grayscale is 400 nm. The two large particles in the center of the image suggest some heterogeneity in the size distribution of the microspheres as these appear significantly larger than the four particles at the top of the image, which each show the apparent size expected for 1.1 μm particles. Figure 2(b) is a near-field transmission image of this film in which the transmission is integrated over the entire bandwidth of the imaging radiation (as depicted in Fig. 1). The transmission (T) contrast (T_{max}-T_{min}/T_{max}) is 0.26 and the PS particles generally appear dark, indicating a decrease in near-field transmission over the particles, although there are also interesting variations in transmssion within the area of the particle (e.g. light bands in the four small particles). Figure 2(c) shows a band ratio absorbance representation of this image data, obtained by taking the ratio of the transmission image resonant with the large C-H aromatic transtion in PS (3020 cm^{-1}) to an off-resonance transmission image (3120 cm^{-1}) and converting to absorbance units. The PS particles in this case appear bright, as is expected given the choice of a PS resonance frequency and the assumption that the principal source of contrast is absorbance. The 1.1 μm particles are easily detected and the spatial resolution as estimated from images of the resolution test sample is nominally 400 nm (~λ/8). Figure 2(d) shows a line scan extracted from Figure 2(c) along the white line wherein the response over the three PS particles is readily apparent. Note that the interparticle separation between the two particles on the right is 1.9 μm and that these particles are resolved well beyond the Rayleigh criteria, clearly demonstrating resolution far exceeding the diffraction limit (λ/2NA ~ 3.3 μm for this collection objective).

The key to the proper interpretation of these images is understanding the source of the image contrast. Several sources are expected to contribute to the measured contrast, including spatial variations in the sample morphology and refractive index. These factors can lead to contrast due to topography, near-field coupling and scattering effects, all of which carry little chemical information. Of course, spatial variations in the the imaginary part of the specimen refractive index gives rise to contrast dependent on the chemical composition of the sample. A strategy for unraveling these multiple, often spatially correlated, sources of image contrast is to measure the wavelength dependence of the contrast and compare it to that expected for possible contrast sources (17). For example, the contrast due to absorbance can be predicted based on the far-field spectra of the material components, while the other mechanisms are expected to have a much weaker dependence on wavelength. Broadband near-field IR microscopy allows for the robust evaluation of the spectral contrast, *i.e.* the variation in image contrast with changes in the imaging radiation frequency. The band ratio image shown in Figure 2(c) is an example of a simple approach to analysis of the spectral contrast, but examination of the spectra recorded at

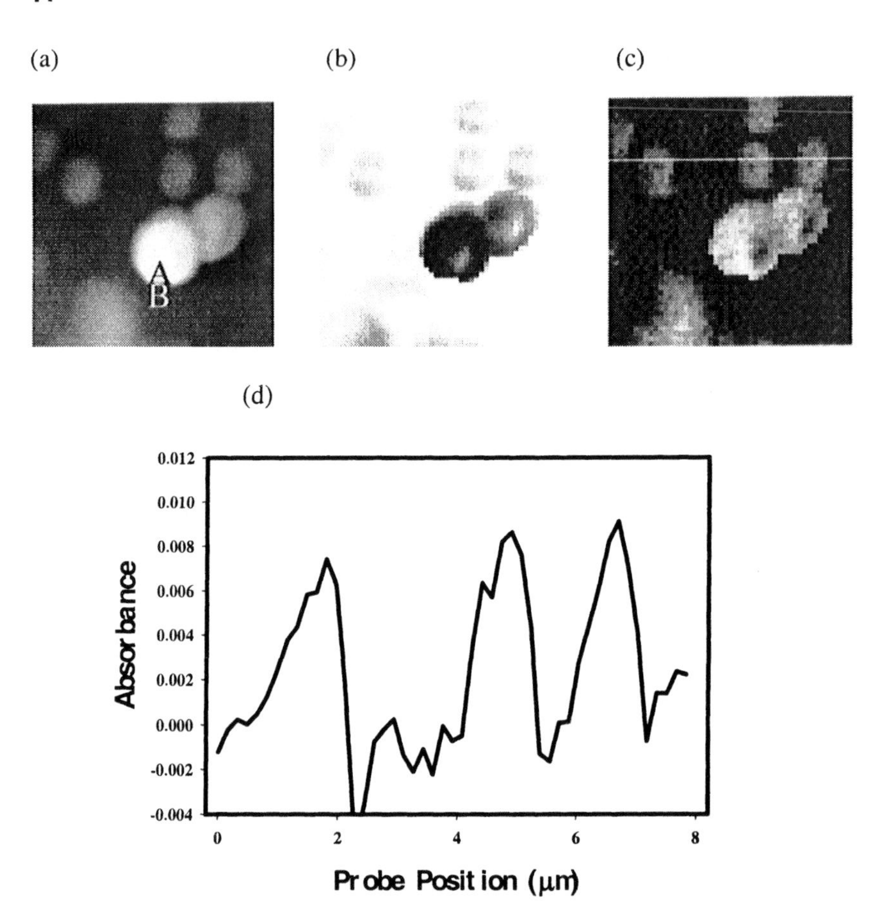

Figure 2. *(a) 8 μm x 8 μm topographic image of 1.1 μm PS microspheres dispersed in a PMMA film. (b) 8 μm x 8 μm CGM near-field, integrated transmission image acquired with the laser frequency centered at 3060 cm^{-1}. (c) 8 μm x 8 μm aromatic C-H stretching band (3020 cm^{-1}), near-field, band ratio image. (d) Line scan extracted along the white line indicated in (c).*

various points on the surface carries even more information. Figure 3 shows the spectral response recorded at two points on the surface, indicated by the letters A (large PS particle) and B (PMMA film) in Figure 2(a). Note that the separation between these pixels is 480 nm, roughly $\lambda/7$. These spectra are normalized to the average spectra recorded over the whole image, in order to remove the lineshape of the laser source (the normalization procedure has been described in detail elsewhere (15)). The B spectrum is roughly equal to unity across the whole spectral range, indicating that the spectral response over this part of the PMMA film is essential identical to that of the average pixel in this image. The A spectrum reveals roughly 20% lower transmission throughout this spectral range with some variation with frequency. In general, this indicates that the predominant source of contrast in this image is not absorbance, as the spectral contrast does not bear any similarity to the known PS spectrum and the low value of transmission is nominally independent of frequency. This suggests that the contrast is due to another source, presumably related to the particle morphology and/or the index contrast between PMMA and PS. The amount of light coupled from the probe out to the detector is a function of the probe-sample separation and the real part of the material refractive index (optical impedance matching (29)). These effects are expected to depend only weakly on frequency, although the real part of the refractive index, n, is not constant in this region due

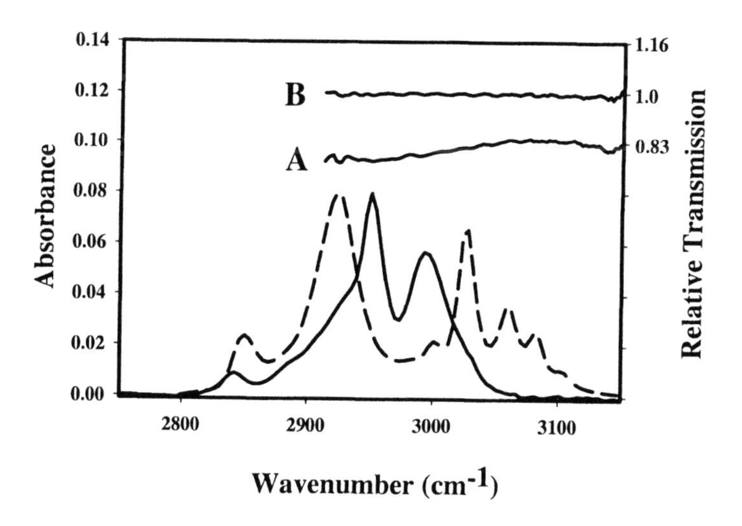

Figure 3. Scaled IR absorbance spectra of PS (dashed line) and PMMA (solid line) in the C-H stretch region along with the normalized near-field transmission spectra extracted at pixels A and B (see Fig. 2(a)).

to the absorbance resonances. This effect is shown in Figure 4, where n is plotted against frequency for PS and PMMA (30). These variations in n are small because the magnitudes of the absorbance resonances are also relatively small and thus significant band-shifting effects, as mentioned in the introduction, are not expected at these frequencies. Scattering effects might also play a role in the spectral contrast, particularly since particle A is nominally 2 μm in diameter, sufficiently large that scattering resonances related to the particle size might play a role, giving rise to spectral variation more complicated than the familiar

Rayleigh dependence for particles much smaller than the wavelength (~ λ^{-4}) (31). Clearly, a detailed understanding of the spectra presented in Figure 3 will require electromagnetic wave propagation calculations, as multiple sources of contrast are relevant and an assignment based solely on the spectra is not possible. That said, it is clear that absorption is not an important source of contrast and interpretation of the band ratio image in Figure 2(c) in terms of absorbance is rather misleading. This highlights the utility of having full spectra at each pixel rather than attempting to interpret images acquired at a select series of frequencies.

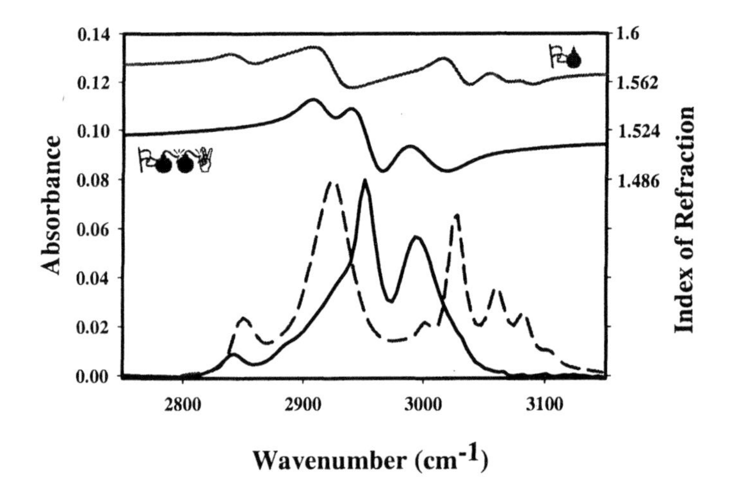

Figure 4. Scaled IR absorbance spectra of PS (dashed line) and PMMA (solid line) in the C-H stretch region along with plots of the real part of the refractive index of these materials versus frequency.

A useful method for analyzing the role of topography induced optical contrast is to compare CGM and CHM images (32). In the first case the probe follows the topography of the surface to the extent possible given its size while in the second mode the separation between the probe and average plane of the sample is kept constant. Figure 5(a) shows an 8 μm x 8 μm CHM near-field transmission image of an individual 1.1 μm diameter PS particle where the optical signal is integrated over the laser bandwidth. The separation between the probe and the PMMA film is maintained constant at approximately 200 nm. In this case, the particle appears bright with transmission contrast of 0.2, indicating increased transmission to the detector as the probe is scanned over the particle, as opposed to the CGM image shown in Figure 2(b) where the particles appear dark. Topographic imaging of this region of the sample indicates that the particle protrudes from the PMMA film by 105 nm. Consequently, the particle couples more light on to the detector than the film, as the field decays very rapidly with increasing distance from the aperture and the probe-sample separation is smaller over the particle than over the film. This near-field coupling effect is certainly one source of the increased transmission over the PS particle. Figure 5(b) is a plot of the normalized spectral response at points A and B (indicated in Figure 5(a)) along with the normalized absorbance spectra of PS and PMMA. The spectra at point B is nominally unity across this region indicating that the response at this pixel is essentially equal to that of the average spectrum used for normalization. The increased transmission at point A is largely independent of frequency, as would be expected for near-field coupling (this will depend on n (16) but this quantity changes only slightly with frequency as shown in Figure 4). Again it is is clear that there is no spectral contrast that is readily identifiable with the far-field absorbance spectrum of PS, indicating that absorbance is not a significant source of contrast in these images. Based on the far-field spectra of PS and PMMA a rough estimate for the expected absorbance contrast between this particle and the film is only a few percent, although even this level of contrast is not evident in the measured spectra. The conclusion that the near-field IR imaging of this composite in the C-H stretching region is dominated by sources of contrast other than absorbance is independent of the scanning mode. This finding has obvious implications for the utility of this technique for chemical imaging, suggesting that for some sample/wavelength combinations, the contrast that carries the desired chemical information will be obscured by other sources of contrast. It is not possible to make specific predictions about samples for which absorbance is likely to be a significant source of contrast; however some general observations can be made. Obviously, absorbance contrast based on strong IR oscillators (*e.g.* imaging of acrylate distribution in a styrene matrix via the carbonyl stretch) will yield images more directly reflective of the distribution of chemical species than those presented in this contribution. The C-H stretch absorbance bands utilized in this work are relatively weak; clearly expansion of wavelength coverage to regions with stronger absorbance bands should be a focus of future work on IR NSOM. However, samples with significant morphological structure and/or components with large disparities in the real part of the refractive index will likely show significant contrast that is unrelated to variations in chemical composition even in spectral regions of significant absorbance.

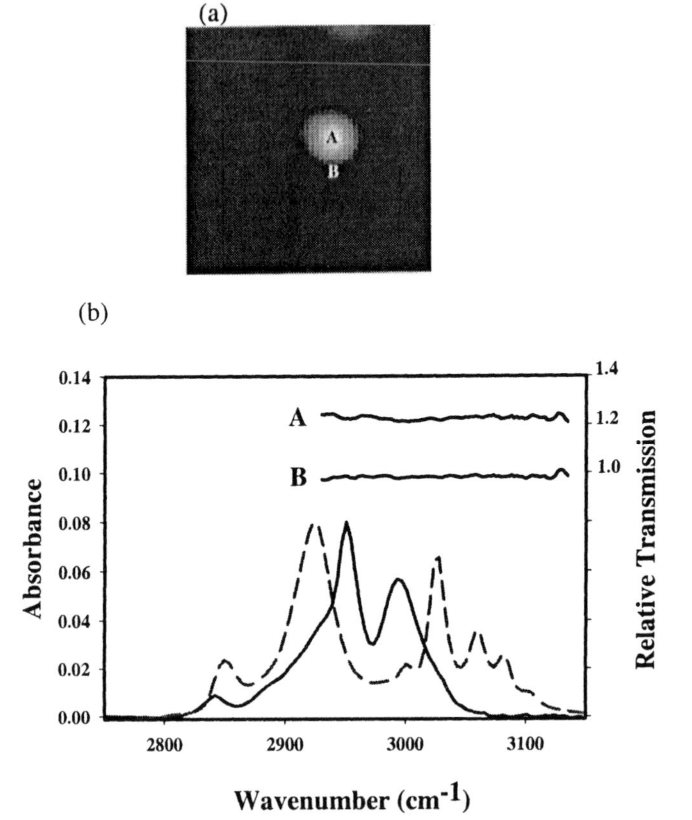

Figure 5. (a) 8μm x 8 μm CHM near-field, integrated transmission image acquired with the laser frequency centered at 3060 cm^{-1}. (b) Scaled IR absorbance spectra of PS (dashed line) and PMMA (solid line) in the C-H stretch region along with the normalized near-field transmission spectra for pixels A and B.

Conclusion

An infrared near-field microscope utilizing a broadband laser source and an fiber-based aperture probe has been used to image a polymer composite consisting of PS particles dispersed in a PMMA matrix. The specimen was imaged in the spectral region encompassing the C-H stretching vibrations of the polymer components. Spectral images acquired in both CGM and CHM imaging modes were reported. The 1.1 μm diameter PS particles were easily distinguished from the PMMA matrix in these high contrast images, with a nominal spatial resolution of 400 nm. Close examination of spectra from individual pixels indicated that absorbance was not an important source of contrast in these images. The weak spectral contrast observed indicated that other sources of image contrast dominated, irrespective of the imaging mode. This data reveals the difficulty associated with interpreting IR near-field images recorded at only a few wavelengths and suggests that, for many samples, the utility of this technique for chemical imaging will be complicated by the presence of multiple contrast mechanisms. Proper image interpretation will require means for identifying and quantifying these multiple types of contrast, including the acquisition of as much spectral data as is practical and the utilization of known sample properties in the application of electromagnetic wave propagation theories to image simulation. Future exploration of chemical imaging with IR NSOM will be most fruitful using appropriate wavelengths for materials with strong IR oscillators, such as the carbonyl stretch modes of acrylate polymers.

References

1. Webster, S.; Batchelder, D. N.; Smith, D. A. *Appl. Phys. Lett.* **1998**, *72*, 1478.
2. Goetz, M.; Drews, D.; Zahn, D. R. T.; Wannemacher, R. *J. Lumin.* **1998**, *76-7*, 306.
3. Jahncke, C. L.; Paesler, M. A.; Hallen, H. D. *Appl. Phys. Lett.* **1995**, *67*, 2483.
4. Stöckle, R. M.; Deckert, V.; Fokas, C.; Zeisel, D.; Zenobi, R. *Vib. Spec.* **2000**, *22*, 39.
5. Emory, S. R.; Nie, S. M. *Anal. Chem.* **1997**, *69*, 2631.
6. Jordan, C. J.; Stranick, S. J.; Cavanagh, R. R.; Richter, L. J.; Chase, D. B. *Surf. Sci.* **1999**, *433*, 48.
7. Narita, Y.; Tadokoro, T.; Ikeda, T.; Saiki, T.; Mononobe, S.; Ohtsu, M. *Appl. Spec.* **1998**, *52*, 1141.
8. Hayazawa, N.; Inouye, Y.; Kawata, S. *J. Microsc.* **1999**, *194*, 472.
9. Piednoir, A.; Creuzet, F. *Micron* **1996**, *27*, 335.
10. Bachelot, R.; Gleyzes, P.; Boccara, A. C. *Opt. Lett.* **1995**, *20*, 1924.
11. Hong, M. K.; Jeung, A. G.; Dokholyan, N. V.; Smith, T. I.; Schwettman, H. A.; Huie, P.; Erramilli, S. *Nuc. Inst. Meth. Phys. Res. B* **1998**, *144*, 246.
12. Knoll, B.; Keilmann, F. *Nature* **1999**, *399*, 134.

13. Schaafsma, D. T.; Mossadegh, R.; Sanghera, J. S.; Aggarwal, I. D.; Gilligan, J. M.; Tolk, N. H.; Luce, M.; Generosi, R.; Perfetti, P.; Cricenti, A.; Margaritondo, G. *Ultramicroscopy* **1999**, *77*, 77.
14. Dragnea, B.; Preusser, J.; Schade, W.; Leone, S. R.; Hinsberg, W. D. *J. Appl. Phys.* **1999**, *86*, 2795.
15. Michaels, C. A.; Stranick, S. J.; Richter, L. J.; Cavanagh, R. R. *J. Appl. Phys.* **2000**, *88*, 4832.
16. Palanker, D. V.; Simanovskii, D. M.; Huie, P.; Smith, T. I. *J. Appl. Phys.* **2000**, *88*, 6808.
17. Michaels, C. A.; Gu, X.; Chase, D. B.; Stranick, S. J. *Appl. Spec.* **2004**, *58*, 257.
18. Akhremitchev, B. B.; Pollack, S.; Walker, G. C. *Langmuir* **2001**, *17*, 2774.
19. Koenig, J. L. *Microspectroscopic Imaging of Polymers*; American Chemical Society: Washington, DC, 1998.
20. Born, M.; Wolf, E. *Principles of Optics*; Cambridge University Press; Cambridge, UK, 1960.
21. Synge, E. H. *Phil. Mag.* **1928**, *6*, 356.
22. Betzig, E.; Finn, P. L.; Weiner, J. S. *Appl. Phys. Lett.* **1992**, *60*, 2484.
23. Stranick, S. J.; Richter, L. J.; Cavanagh, R. R. *J. Vac. Sci. Tech. B* **1998**, *16*, 1948.
24. Betzig, E.; Trautman, J.K. *Science* **1992**, *257*, 189.
25. Schaller, R. D.; Saykally, R. J. *Langmuir* **2001**, *17*, 2055.
26. Stöckle, R.; Fokas, C.; Deckert, V.; Zenobi, R.; Sick, B.; Hecht, B.; Wild, U. P. *Appl. Phys. Lett.* **1999**, *75*, 160.
27. Unger, M. A.; Kossakovski, D. A.; Kongovi, R.; Beauchamp, J. L.; Baldeschwieler, J. D.; Palanker, D. V. *Rev. Sci. Instrum.* **1998**, *69*, 2988.
28. Certain commercial equipment, instruments, or materials are identified in this paper to specify adequately the experimental procedure. In no case does such identification imply recommendation or endorsement by the National Institute of Standards and Technology, nor does it imply that the materials or equipment identified are necessarily the best available for the purpose.
29. Campillo, A. L.; Hsu, J. W. P.; Bryant, G. W. *Opt. Lett.* **2002**, *27*, 415.
30. Woolam, J. A. personal communication.
31. Bohren, C. F.; Huffman, D. R. *Absorption and Scattering of Light by Small Particles*; John Wiley & Sons; New York, NY, 1983.
32. Hecht, B.; Bielefeldt, H.; Inouye, Y.; Pohl, D. W.; Novotny, L. *J. Appl. Phys.* **1997**, *81*, 2492.

Chapter 5

Apertureless Scanning Near-Field IR Microscopy for Chemical Imaging of Thin Films

Boris B. Akhremitchev[1], Larissa Stebounova[2], Yujie Sun[2], and Gilbert C. Walker[2]

[1]Department of Chemistry, Duke University, Durham, NC 27708
[2]Department of Chemistry, University of Pittsburgh, Pittsburgh, PA 15260

The sample's absorption of infrared radiation is used for chemical mapping of thin films using apertureless near-field infrared microscopy. An oscillating metallic probe modulates the scattering of the infrared radiation at the surface. The near-field signal is registered by demodulation of the scattered signal at twice the probe's oscillation frequency. An apparent image contrast contains contributions from both the topography and the optical properties of the sample. We demonstrate that for a sufficiently thin sample the topography contribution can be small enough to detect the chemical variations unambiguously with a spatial resolution of ca 200 nm. The signal dependence on probe-sample separation is well described by employing a quasi-electrostatic model. This model includes scattering from two sources of different sizes. We believe that these scattering sources correspond to the different parts of the probe.

Introduction

Spatial resolution of far-field microscopies is limited to approximately λ/2 due to decay of radiation components with high spatial frequency. This limits spatial resolution of chemical mapping with infrared radiation to ~5 μm. Near-field techniques that probe the electromagnetic field at the surface of the sample remove such limitation. These techniques use scanning probes that convert evanescent field components at the surface of the sample into propagating components. Detection of the propagating electromagnetic field that results from such scattering provides information about local optical properties of the sample. Near-field microscopes can be subdivided into two types. The first type is an aperture-based approach that uses a sub-wavelength aperture at the end of tapered fiber (Figure 1a). The aperture is placed near the sample's surface as a light source, and the electromagnetic (EM) radiation that is scattered, transmitted or emitted by the sample is detected (*1*). This aperture probe approach typically provides ~λ/20 resolution in the visible part of the EM spectrum (*1*) and at best a resolution of λ/10 in the infrared region (2) (λ is wavelength of light). The second approach, known as "apertureless" or "scattering" approach, shown in Figure 1b, employs a nanometer-sized probe as a local scatterer of EM radiation in the vicinity of the sample surface (*3, 4, 5, 6*). Apertureless imaging provides higher resolution than the apertured probe approach, with reported resolution exceeding λ/100 in the visible (7) and λ/300 in the infrared (8).

Interpretation of the scattering data is not straightforward. Topography of the sample and background scattering both contribute to the signal. In this paper, we describe two implementations of the apertureless near-field

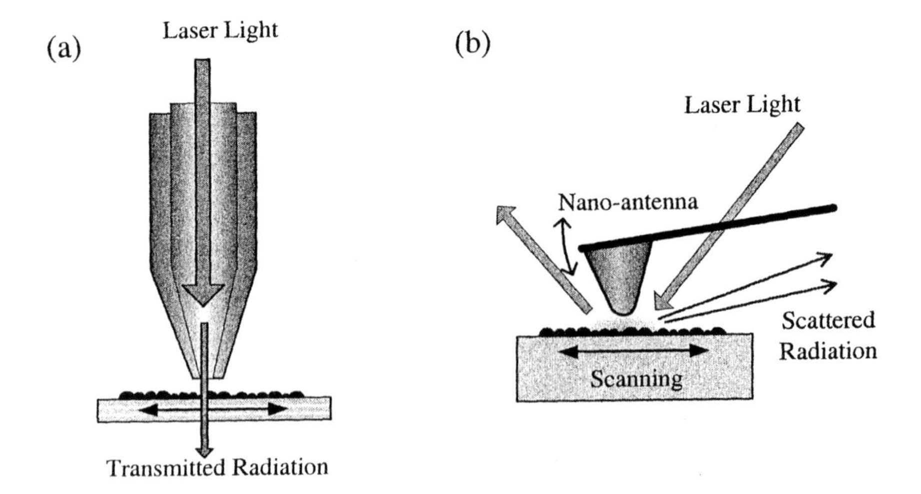

Figure 1. Schematic diagrams showing arrangements of a) aperture-based and b) apertureless near-field instruments.

microscope, along with approaches for artifact removal, followed by the analysis of the scattered signal-distance dependence. In the first approach (*9*), the topography-related artifacts are removed by employing the signal interpolation procedure to a plane directly above the tallest feature of sample. The second approach (*10*) removes the contribution of the background scattering by the second harmonic demodulation of the scattered signal.

Detection of chemical contrast in the presence of large topographic artifact

Experimental setup

The near-field microscope was developed from a commercial atomic force microscope (Multimode AFM, Digital Instruments, Santa Barbara, CA). Figure 2 shows the scheme of the apparatus. The infrared light emitted by a carbon dioxide laser was focused onto the end of a tungsten-coated cantilever probe perpendicular to the long axis of the probe. We use *p*-polarized light with a 73° angle of incidence. The spot size is approximately 100 μm and the radiation power was 20-100 mW. Oscillation of the probe (100-200 nm total amplitude) was excited by mechanical driving the probe near its resonance frequency.

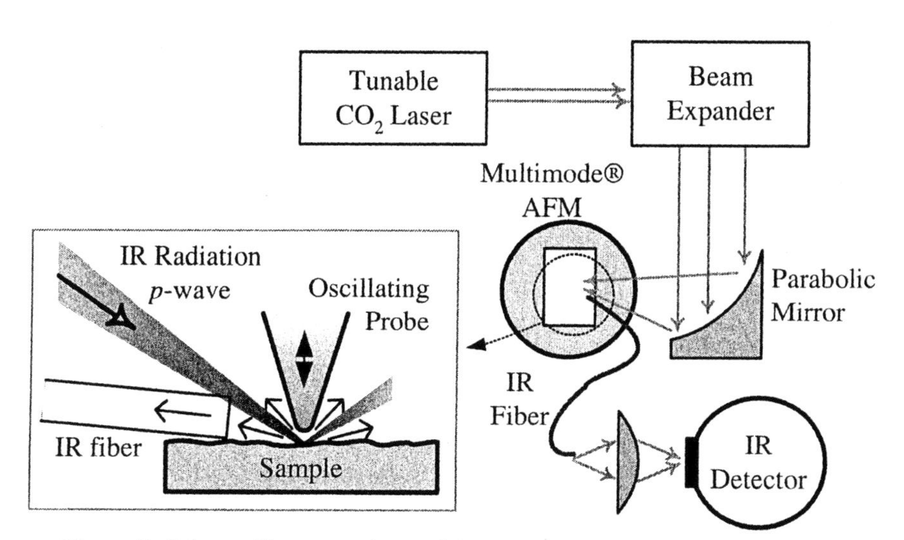

Figure 2. Scheme illustrates the modification of the tapping mode scanning AFM into an apertureless near-field scanning infrared microscope.

Imaging of the surface of the sample in intermittent-contact (tapping) mode was performed simultaneously with the detection of the scattered infrared signal. Scattered light was collected by an infrared-transparent multimode optical fiber (500 μm chalcogenide glass core) that was placed near the probe (~ 0.5 mm away). Light collected by the fiber was detected with a MCT infrared detector (Graseby Infrared, Orlando, FL). The electrical signal from the detector was amplified by a lock-in amplifier (SR844, Stanford Research Systems, Sunnyvale, CA) at the frequency of the cantilever oscillation and collected by a computer simultaneously with AFM data. The use of an oscillating cantilever eliminated the DC background signal (*11*) and provided the benefits of sensitive lock-in detection.

The polymer film sample was prepared by spin casting a 1.5 mg/ml solution of polystyrene-polydimethylsiloxane diblock polymer (*12*) in CH_2Cl_2 at 3000 RPM using a Headway Research ED101 photo resist spinner. A gold-coated microscope cover slip was used as the substrate.

Results

Figure 3 shows topography and near-field data collected with the near-field microscope (*9*). Topographic images of the sample indicate that ring-like structures have formed on the surface of the sample. A scattered infrared signal modulated by the oscillating probe (henceforth "infrared signal") was collected at a frequency of a CO_2 laser where the PDMS component of the polymer demonstrates a noticeable absorption, in contrast to the PS block (982 cm^{-1}). Therefore, the observed infrared signal contrast (right panel) might be taken to indicate a chemical composition difference at the sample's surface. On the other hand, a high correlation between the two images can result from constant-gap scanning of non-flat objects (*13, 14*). Topographic contributions into the infrared signal can be excluded by scanning above the surface *without* following the topography. This is achieved by collecting a number of infrared signal maps at different mean probe-sample separation distances and interpolating the volume of data to a plane that is located just above the tallest feature of the

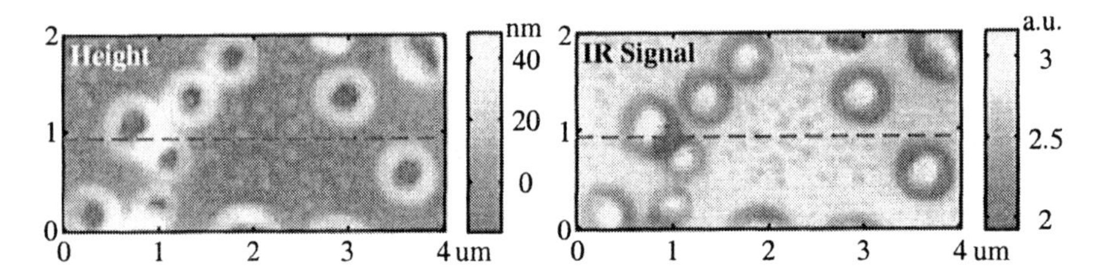

Figure 3. Tapping mode AFM maps of the PS-PDMS sample topography (left panel) and simultaneously collected infrared signal (right panel).

sample to produce a constant height image. Once the constant height plane is selected, the corresponding probe-sample separation can be calculated for each point in the image. This separation value is used to interpolate the IR signal measured at different gap values, producing the constant height image.

Panel A of Figure 4 shows seven line plots collected at different probe-sample gaps that are indicated below each line. Lines are shifted in the graph for clarity. Corresponding topography of the sample is indicated in Panel C. The plane of constant height is fixed parallel to the flat areas of the sample (areas without the ring-like structures). Scanning position is indicated in Figure 3 with a dashed line. Panel B shows the interpolated infrared signal. Comparison between data in Panels A and B indicates that the interpolation procedure eliminates the decrease in the infrared signal that is observed above the rings. Panels D and E of Figure 4 compare the constant gap and constant height maps. It can be noticed that the infrared signal contrast that remains after interpolation is associated with small topographic features that can be seen in the middle of height image of Figure 3. These features exhibit negative contrast arising from local variations of optical properties, namely due to light absorption by the PDMS block of the polymer. This suggestion is supported by the known surface segregation in spin-cast films of PS-PDMS diblock copolymers. (*15*) The surface coverage by PDMS is not complete for the 0.3 mole fraction PDMS block, which might explain observed patches. (*15*)

The indicated procedure can be used to remove surface-following artifacts.

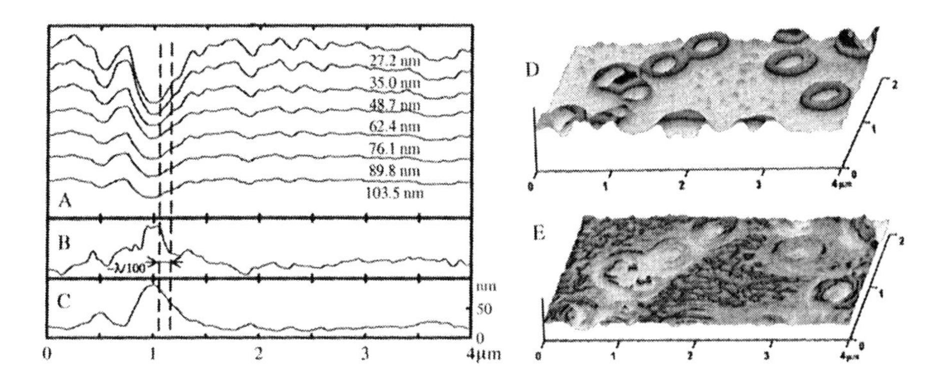

Figure 4. This figure shows the results of interpolation procedure. Panel A shows the constant gap infrared signal collected at different heights along the line that is indicated in Figure 3. Panel B shows interpolated constant height infrared signal, and Panel C shows the topography of the sample along this line. Panels D and E show constant gap and interpolated constant height signals respectively. The constant height signal is very different from the constant gap signal indicating that the contrast observed during constant gap scanning comes mostly from the topography following and not from the difference in chemical composition of the sample.

The disadvantage of this approach is that it requires collection of several images. This is a lengthy procedure that can be used only with low lateral drift of the AFM scanner. Significant drawbacks of the presented instrument are that the instrument was surprisingly difficult to align in order to achieve high signal to noise ratio in the detected infrared signal and that the infrared signal is not always surface-specific. Figure 5 shows the simultaneously collected probe oscillation amplitude and infrared signals. The infrared signals are demodulated from the detected infrared intensity at the frequency of the probe oscillation and at the frequency that is twice the probe oscillation frequency.

The amplitude of the probe oscillation decreases at separation less than approximately 30 nm due to probe-sample interaction. A decrease in the oscillation amplitude results in a corresponding decrease in the infrared signal. Close to the surface of the sample the infrared signal decreases as expected for a surface specific near-field signal; whereas at distances of approximately 75 nm, the signal (thick line in the bottom panel) starts to increase. Such an increase in the signal is probably caused by the interference of infrared radiation from different sources (not just part of the probe that is close to the surface). Difficulty in tuning can be understood by considering that the magnitude of the infrared radiation detected from the small scattered probe (with size of 100 nm) is considerably less than the power of infrared radiation that is necessary to achieve detected infrared signal. This discrepancy can be explained by noting that the high level of demodulated signal coincided with the high level of dc

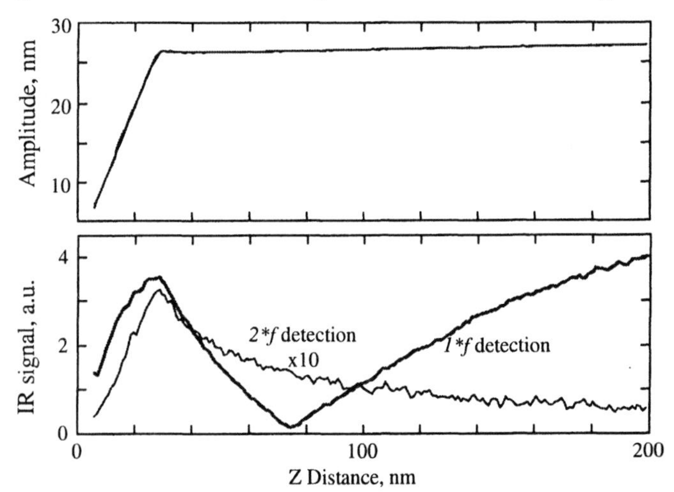

*Figure 5. The top panel shows the dependence of the amplitude of the vertical probe oscillation on the separation from the surface. The bottom panel shows IR signals demodulated at the frequency of the probe oscillation (**f**) and twice the oscillation frequency (**2f**, see later in the text). The signal at **2f** is multiplied by 10. The interferometric feature present in the **f**-signal is missing from the **2f**-signal.*

infrared signal. Therefore, homodyne amplification of infrared radiation scattered by the probe can occur at the detector. The above experimental setup does not provide a means for simultaneous optical alignment of the scattered radiation and the reference field, therefore a high signal to noise ratio was difficult to achieve. The next section describes an approach that eliminates these problems and is capable of sub-monolayer sensitive chemical imaging.

Monolayer-sensitive infrared imaging

Detection of the surface specific signal

The rapid decrease of the evanescent components of EM radiation with increasing distance from the surface can be utilized to achieve a surface-specific signal detection (*10* and references therein). This can be accomplished by detection of the scattered signal at twice the frequency of the probe's oscillation. It has been established that probe oscillation, upon approach toward the surface, remains harmonic (*16*). The presence of the second harmonic in the scattered signal indicates strong non-linearity in the near-field signal upon approach to the surface. Figure 5 shows both the distance dependence of the cantilever oscillation amplitude (top panel) and the scattered IR signals (lower panel). The signals that are collected at one and two times the frequency of the cantilever oscillation are called the ***f***- and ***2f***-signals, respectively. In the region where the cantilever oscillation amplitude is relatively constant (30-200 nm from the surface), the ***2f***-signal decreases monotonically with increasing tip-surface separation, while the ***f***-signal does not. Such non-monotonic dependencies can produce artifacts during imaging. At distances less than ~30 nm the amplitudes of both signals decay due to damping of the probe oscillation by the surface. It can be noted that the ***2f***-signal is about an order of magnitude weaker than the ***f***-signal. Similarly, it was possible to detect the ***3f***-signal, but it was considerably weaker than the ***2f***-signal (data not shown).

Homodyne amplification of scattered signal

The detected infrared signal is proportional to the square of the electric field at the detector.

$$I_t = \left\langle (E_{sc} + E_r)^2 \right\rangle = \left\langle E_{sc}^2 \right\rangle + \left\langle E_r^2 \right\rangle + 2\left\langle E_{sc} E_r \right\rangle \qquad (1)$$

The weak scattered signal E_{sc} can be amplified by a much stronger reference field E_r. We can substitute $E_r = A_2 \cos(\omega t)$ for the reference field (ω is the frequency of EM radiation) and $E_{sc} = A_1(1 + C_1\cos(f \cdot t + \Phi) + C_2\cos(2f \cdot t + \Phi))\cos(\omega t + \varphi)$ for the scattered field into Equation 1. The expression for the scattered field includes electric field modulation by the probe motion at ***f*** and ***2f*** and a phase offset φ between scattered and reference fields. The detected intensity demodulated at ***2f*** will be proportional to $I(2f) = A_1^2(\frac{C_1^2}{4} + C_2) + A_1A_2C_2\cos\varphi$. In this equation, the most significant term is the second term containing the amplitude of the reference field (A_2). Therefore, providing a separate reference field in the measurements can significantly increase the signal demodulated at twice the cantilever oscillation frequency.

A reference arm was added to the instrument in order to facilitate a homodyne signal amplification. The variable length reference arm utilized a partial reflector to direct the reference field onto the IR detector. The length of this arm was adjusted to maximize the demodulated signal (*17*).

Figure 6 shows data that were collected with and without the reference field. The top panel shows the amplitude of the probe and the bottom panel shows the infrared signal collected with and without the reference beam. No instrument adjustment was performed between these two measurements. There is no signal until the reference field is provided. Often some ***2f*** signal was detected without the reference beam. In this case, background light scattered from the probe and the sample acts as a reference field.

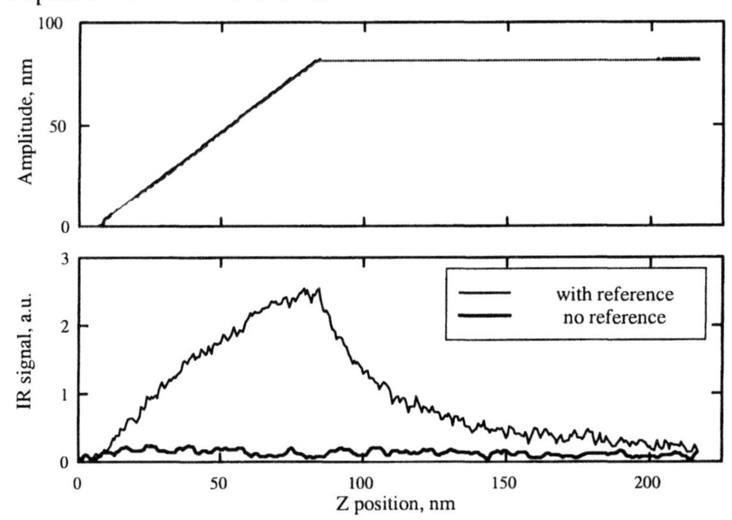

Figure 6. The top panel shows the dependence of the amplitude of the probe oscillation on the separation from the surface. The bottom panel shows IR signals demodulated at twice the oscillation frequency that were detected with (thin line) and without (thick line) the reference beam.

Detection of DNA monolayer

The discussed strategies have been implemented in the improved near-field microscope. Instrument sensitivity and tunability were improved by using a reflective infrared objective (0.28 NA, Coherent Inc., Auburn, CA) and by employing a back-scattering configuration (*10, 17*).

The patterned DNA samples were prepared by creating a pattern of 1-hexadecanethiol with micro-contact printing (*18*) and consequent reaction of unprotected areas with thiolated DNA solution (0.25 mM solution of 24-Cytosine single strand 3'-tholated DNA oligomer in deionized water). After incubation, substrates were cleaned of the excess DNA by sonication in deionized water and dried in a stream of nitrogen gas.

Figure 7 shows two 20 μm x 20 μm images. Panel A shows height image and Panel B shows the simultaneously collected near-field image. Infrared radiation with frequency 980 cm^{-1} was scattered by the commercial platinum-coated probe. Probe was oscillating with a ~100 nm peak-to-peak amplitude at ~133 kHz. Infrared radiation corresponds to the phosphate IR absorption band of DNA. (*19*) Lock-in detection at ***2f*** (~266 kHz) was accomplished with a 3 ms time constant. The scan rate was 0.3 Hz. The mean near-field ***2f***-signal collected during scanning corresponds to ~0.5 nW of infrared radiation power on the detector, modulated at ***2f***. The striped pattern is visible in the topography image, although clear edges of the DNA stripes are not readily discernable in this image. The near-field image shows a periodic variation corresponding to the periodicity and structure of the stamp that was used in the sample's preparation. Here the near-field map does not directly correlate with the topographic map; the

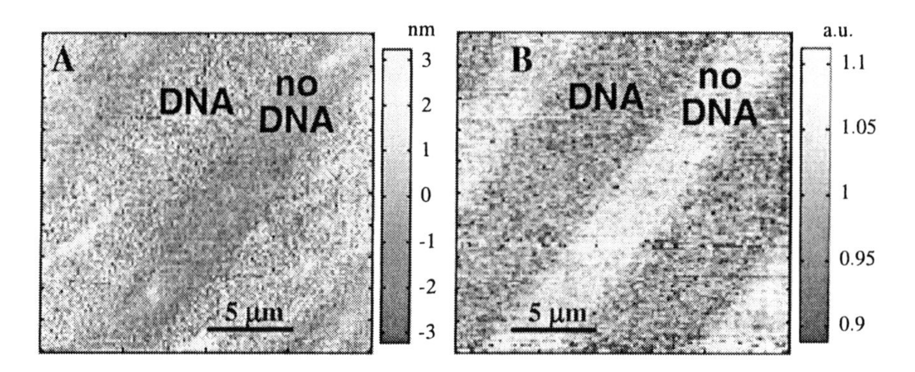

Figure 7. Panel A shows the topography of the striped DNA- hexadecanethiol sample, Panel B shows the **2f** *infrared signal. The 10 μm stripe pattern can be easily seen in the IR signal image; DNA regions are darker than alkanethiol regions. Several height features discernable in the topographic image are not coupled to IR signal.*

normalized cross-correlation coefficient between images is –0.15. Due to the low correlation between the topography and the near-field signal in Figure 7, we conclude that the near-field signal is not topography-coupled. The edge contrast in the near-field signal indicates that the spatial resolution of the instrument is at least 200 nm (10-90% transition). It is believed that this edge sharpness is affected by the sample preparation procedure, and the resolution achievable with this instrument is higher and limited by the size of the probe (~50 nm) and the mean probe-sample distance (50 nm).

The near-field absorption spectrum can be constructed by comparing the average infrared signal in 1-hexadecanethiol and DNA coated areas. Figure 8 shows such spectrum that was collected using data from eleven different wavelengths (triangles). This spectrum is compared to the far-field infrared spectrum that was collected using similarly prepared monolayer sample (dots) at 70° angle of incidence. The near-field spectrum was created by averaging many data files. The width of the phosphate band exceeded the tunability of carbon dioxide laser precluding measurements on both sides of this absorption band. The spectral dependence of the near-field signal further confirms that the observed contrast is caused by variation in chemical composition across the surface.

We also note that the near-field absorption follows the far-field absorption and not the variation in the real part of index of refraction. The expected variation in the real part of refraction index is indicated by thin line in Figure 8.

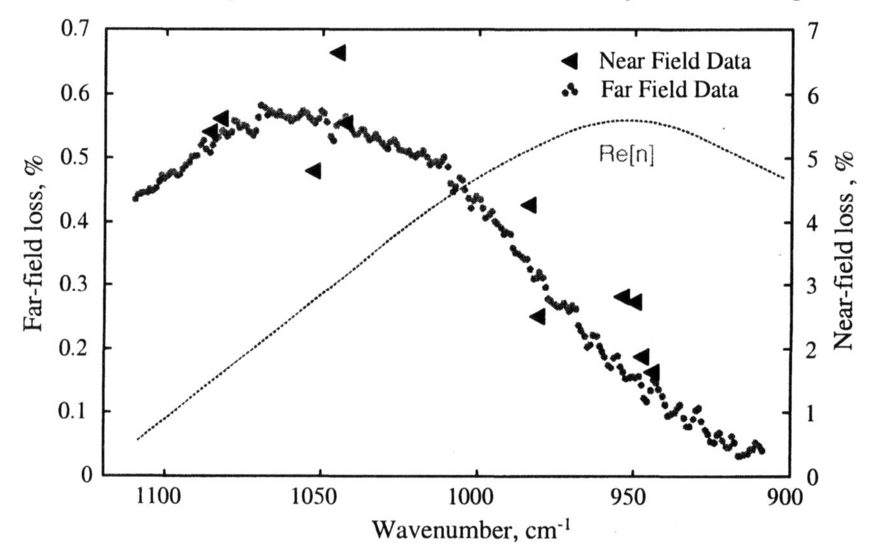

Figure 8. Near-field and far-field absorption spectra of DNA sample in the region of phosphate absorption. Line shows the expected variation in the real part of refraction index of DNA.

Near-field absorption change across the observed part of the phosphate band exceeds the far-field absorption change by almost an order of magnitude (4% change vs. 0.5% change). This enhancement of infrared IR absorption could be related to the well-known phenomenon of the surface-enhanced infrared absorption. (*20*) In our case the increase of absorption could be explained by an electric field enhancement (*21, 22*) under the metallic tip.

The height contrast between 1-hexadecanethiol coated and DNA coated areas is considerably less than expected for dense monolayer coverage of DNA. Observed height of DNA monolayer is ~3 nm and for the spot with diameter of 200 nm; this corresponds to $\sim 10^{-18}$ moles of phosphate groups. It is possible that the number of absorbing phosphate groups is much less due to DNA chain orientation, and that our actual sensitivity is higher.

Distance dependence of near-field signal

Distance dependence of the near-field signal can be used to test existing theoretical models. One model (*23*) uses a dipolar interaction in the electrostatic limit. This model can be simplified when used for metals in the infrared region since the real part of permittivity at 1000 cm^{-1} is large. Then the effective polarizability α_{eff} of a spherical metal probe above the gold surface is

$$\alpha_{eff} = \frac{8\pi a^3}{1 - \frac{a^3}{4(a+z)^3}} \tag{2}$$

where a is the radius of the probe, and z is the probe-sample separation. Intensity of the scattered radiation is proportional to the square of the polarizability (*24*), and the homodyne-detected signal is proportional to the absolute value of polarizability.

During the oscillating of the probe α_{eff} changes significantly (for probe with 50 nm radius oscillating with 100 nm peak-to-peak amplitude α_{eff} changes from $0.42 \cdot 10^{-20}$ m^3 to $0.32 \cdot 10^{-20}$ m^3). Therefore, we cannot assume that the ***2f*** lock-in signal is proportional to the second derivative of α_{eff}. Thus the lock-in detection of the second harmonic signal was modeled by calculation of numeric sine and cosine transforms of polarizability *vs.* time dependence at the twice the frequency of the oscillating cantilever. (*17, 25*) Modulation of the probe position is a harmonic function with $z(t) = z_{mean} + A \cdot \cos(2\pi\nu t)$ where z_{mean} is the mean probe-surface separation, A is the amplitude of the probe motion and ν is frequency of oscillation. Amplitude of the probe oscillation was measured simultaneously with the near-field signal measurements. The probe size was a fit parameter.

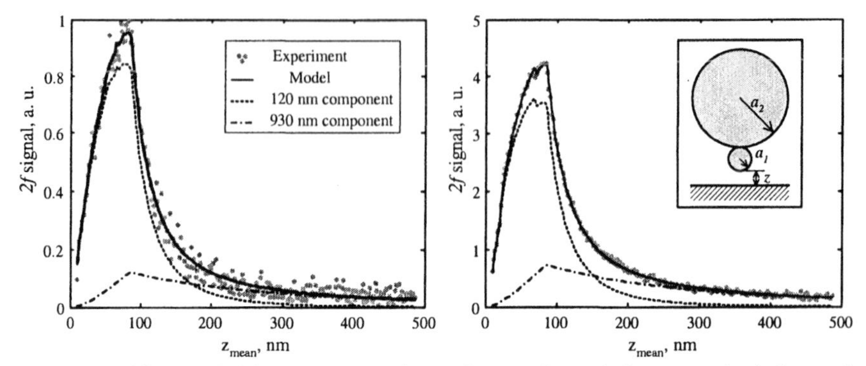

Figure 9. 2f near-field vs. distance dependence. Signal shown in the left panel was collected without the reference beam and in the right panel with the reference beam.

This model had to be further modified in order to fit the observed signal-distance dependence. (*17*) We had to assume that scattering comes from two sources with different size. This modification more accurately describes the true shape of the probe used in the measurements. Figure 9 shows the results of applying this model to two near-field data sets that were collected with (right panel) and without (left panel) the reference beam. The insert in the right panel shows the probe-sample geometry that was assumed in the data reduction. Both sets of data were fit using the same size parameters of the scatterers, 120 nm and 930 nm. The plots show experimental data with grey dots and calculated dependencies as solid lines. Signals due to individual scatterers are indicated with dashed lines. The close fit of both data sets indicate that without the use of reference beam, near-field signal remains amplified by the back-scattered radiation.

Expected signal for semi-infinite polymer sample

The quasi-electrostatic model (*23*) with explicit calculation of the lock-in signal (*17*) can be used to predict the near-field spectra of thick polymer samples. Values for the real and imaginary parts of permittivity for various polymers can be found in the literature. (*26*) Figure 10 shows the expected near-field spectrum (gray thick line) above the semi-infinite polystyrene sample that could be collected by detecting a ***2f*** scattering signal from the metallic probe with 100 nm radius of curvature oscillating at the surface with amplitude of 70 nm. The wavelength dependences of the real (solid line) and imaginary part (dashed line, shifted for clarity) of the refraction index are indicated in the graph.

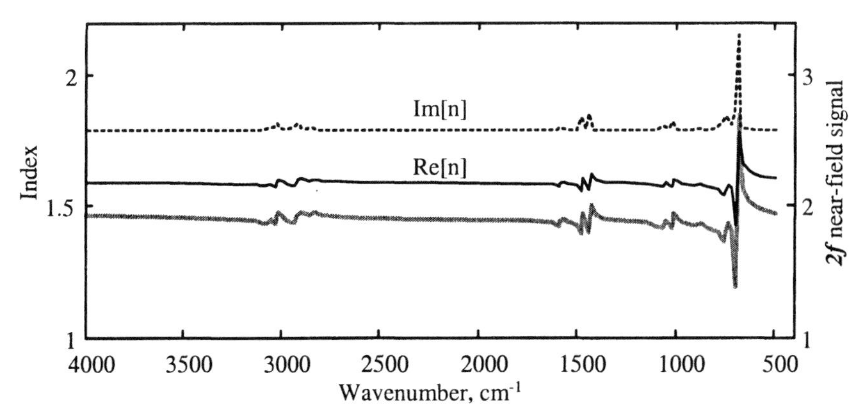

*Figure 10. Expected wavelength dependence of **2f** near-field signal (gray line), real part (solid line) and imaginary part (dashed line) of the refraction index. Imaginary part of the index is shifted and scaled for clarity.*

At the strongest absorption line the near-field signal is modulated my more than 50% of its magnitude (at ~700 cm^{-1}). The near-field follows mostly the refractive index dependence rather than absorption. Both parts of complex refraction index contribute to the signal; the contribution of the real part exceeds the contribution of the imaginary part by almost an order of magnitude. Contribution by the imaginary part of the index becomes comparable only for very absorbing polymers such as poly(perfluoropropylene oxide).

Additional measurements are necessary to validate the above prediction. Though the quasi-electrostatic model predicts sufficiently the distance dependence of the near-field signal for metal-coated probe above the gold surface, further developments are necessary to model the wavelength dependence of the scattered near-field signal for thin samples.

References

1. Hecht, B.; Sick, B.; Wild, U. P.; Deckerd, V.; Zenobi, R.; F. Martin, O. J.; Pohl, D. W. *J. Chem. Phys.* **2000**, *112*, 7761.
2. Dragnea, B.; Preusser, J.; Schade, W.; Leone, S. R.; Hinsberg, W. D. *J. App. Phys.* **1999**, *86*, 2795.
3. Wessel, J. *J. Opt. Soc. Am. B* **1985**, *2*, 1538.
4. F. Zenhausern, M. P. O'Boyle, H. K. Wickramasinghe *Appl. Phys. Lett.* **1994**, *65*, 1623.
5. Inouye, Y.; Kawata, S. *Optics Letters* **1994**, *19*, 159.

6. Bachelot, R.; Gleyzes, P.; Boccara, A. C. *Microsc. Microanal. M.* **1994**, *5*, 389.
7. Zenhausern, F.; Martin, Y.; Wickramasinghe, H. K. *Science* **1995**, *269*, 1083. Hamann, H. F.; Gallagher, A.; Nesbitt, D. J. *Appl. Phys. Lett.* **1998**, *73*, 1469. Hubert, C.; Levy, J. *Appl. Phys. Lett.* **1998**, *73*, 3229. Gresillon, S.; Cory, H.; Rivoal, J. C.; Boccara, A. C. *J. Opt. A: Pure Appl. Opt.* **1999**, *1*, 178.
8. Lahrech, A. ; Bachelot, R.; Gleyzes, P.; Boccara, A. C. *Appl. Phys. Lett.*, **1997**, *71*, 575. Knoll, B.; Keilmann, F. *J. Microscopy* **1999**, 194, 512.
9. Akhremitchev, B. B; Walker G. C. *Langmuir* **2001**, *17*, 2774-2781.
10. Akhremitchev, B. B.; Sun, Y.; Stebounova, L.; Walker, G. C. *Langmuir* **2002**, *18*, 5325-5328.
11. Adam, P. M.; Royer, P.; Laddada, R.; Bijeon, J. L. *Ultramicroscopy* **1998**, *71*, 327. Yamaguchi, M.; Sasaki, Y.; Sasaki, H.; Konada, T.; Horikawa, Y.; Ebina, A.; Umezawa, T.; Horiguchi, T. *J. Microscopy* **1999**, *194*, 552.
12. Pollack, S. K.; Singer, D. U.; Morgan, A. M. *Polymer Preprints* **1999**, *40*, 370.
13. Hecht, B.; Bielefeldt, H.; Inouye, Y.; Pohl, D. W.; Novotny, L. *J. Appl. Phys.* **1997**, *81*, 2492.
14. Knoll, B.; Keilmann, F.; Kramer, A.; Guckenberger, R. *Appl. Phys. Lett.* **1997**, *70*, 2667.
15. Chen, X; Gardella, J. A. *Macromolecules* **1992**, *25*, 6621-6630.
16. Anczykowski, B.; Gotsmann, B.; Fuchs, H.; Cleveland, J. P.; Elings, V. B. *Appl. Surf. Sci.* **1999**, *140*, 376–382.
17. Stebounova, L.; Akhremitchev, B. B.; Walker, G. C. *Rev. Sci. Inst.* **2003**, *74*, 3670-3674.
18. Xia, Y.; Whitesides, G. M. *Annu. Rev. Mater. Sci.* **1998,** *28*, 153–184.
19. Taillandier, E.; Liquier, J. *Methods Enzymol.* **1992**, *211*, 307-335.
20. Osawa, M. *Top. Appl. Phys.* **2001**, *81*, 163-187.
21. Martin, O. J. F.; Girard, C. *Appl. Phys. Lett.* **1997**, *70*, 705-707.
22. Larsen, R. E.; Metiu, H. *J. Chem. Phys.* **2001**, *114*, 6851-6860.
23. Knoll, B.; Keilmann, F. *Opt. Commun.* **2000**, *182*, 321-328.
24. Bohren, C. F.; Huffman, D. R. *Absorption and scattering of light by small particles* ;Wiley & Sons, New York, 1983.
25. Laddada, R.; Benrezzak, S.; Adam, P. M.; Viardot, G.; Bijeon, J. L.; Royer, P. *Eur. Phys. J.-Appl. Phys.* **1999**, *6*, 171-178.
26. Pacansky, J.; England, C.; Waltman, R. J. *J. Polym. Sci. B* **1987**, *25*, 901-933.

Chapter 6

Measurement of the Local Diattenuation and Retardance of Thin Polymer Films Using Near-Field Polarimetry

Lori S. Goldner[1], Michael J. Fasolka[2], and Scott N. Goldie[1,3]

[1]Optical Technology Division, Physics Laboratory and [2]Polymers Division, Materials Science and Engineering Laboratory, National Institute of Standards and Technology, 100 Bureau Drive, Gaithersburg, MD 20899
[3]Current address: Shire Laboratories, Rockville Pike, Rockville, MD 20850

Near-field scanning optical microscopy and Fourier analysis polarimetry are combined to obtain quantitative maps of the local retardance, (resulting from strain or crystalline birefringence), fast axis orientation, diattenuation and diattenuating axis orientation in nanostructured polymer thin films. Lateral resolution of 50 nm with retardance sensitivity as small as 1 mrad has been demonstrated in images of isotactic PS crystallites and diblock copolymer morphologies.

Many techniques are available to study strain and order in bulk polymers, but observing local structure in thin films is complicated by the difficulty of sample preparation, small contrast requiring high sensitivity, and a need for high spatial resolution. Ellipsometry and polarimetry provide high sensitivity but lack spatial resolution necessary for characterizing sub-micron structures. Transmission electron microscopy offers high spatial resolution but requires difficult sample preparations. Aperture-based near-field scanning optical microscopy (*1, 2*) (NSOM) provides a means to measure and view structures as small as 20 nm in size, but most often gives only qualitative images. Here we demonstrate a quantitative extension of NSOM that permits high sensitivity measurements of local optical properties.

Simple static near-field polarimetry (NFP), where a specific polarization state of light is used to excite the sample and another polarization state is detected, has been discussed by many authors (*3-6*) in many different contexts, including investigations of small metal structures (*3, 6*), magnetic films (*3, 7, 8*), lipid films (*9*), J-aggregates (*10*), conjugated polymers (*11-14*) and liquid crystal (LC) droplet structure, dynamics (*15*) and birefringence (*15, 16*). The introduction of polarization modulation techniques to NSOM (*4*) increased both the quality and the information content of polarimetric images. For example, Ade *et al.* (*17*) used a modulating analyzer to improve on static polarimetry measurements; by modulating the analyzer one can obtain both parallel- and crossed-polarizer images simultaneously. Polarization modulation (PM) of the excitation light was used in conjunction with NSOM by Higgins *et al.* to measure the orientation of mesoscopic crystals (*18*), and by the group of H. Heinzelmann to study magnetic materials and liquid crystals (*19-21*). Correcting for tip diattenuation (*22*), Tan *et al.* (*23*) and Wei *et al.* (*24*) used the same technique to study diattenuation of conjugated polymers. Other PM schemes in NSOM can be found (*25-30*). Our scheme most closely resembles that of J.W.P Hsu, whose group studied the local retardance of semiconductors (*31*).

Here we describe our implementation and application of NFP to thin polymer films. The analysis required for measurements of local retardance and diattenuation, including the orientation of corresponding optical axes, is discussed, and the importance of accurately accounting for the diattenuation and residual birefringence of the near-field probe is stressed. Systems studied include (1) ultrahigh molecular weight block copolymers (*32-34*) which microphase separate (*35*) to form domains patterned on a ≈100 nm length scale and (2) polymer crystallites grown in thin films of isotactic polystyrene.

Polarimetric near-field scanning optical microscope

Excellent reviews of NSOM have been written by Pohl (*36*), Betzig (*37*), and more recently Dunn (*38*). A schematic of our NSOM polarimeter is shown in Fig. 1 and described in more detail in Ref. (*39*). A polarizer prepares linearly polarized light at 0°. The polarization generator consists of a Hinds Instruments photoeleastic modulator (PEM) tuned to a nominal modulation frequency $\omega = 50$ kHz, and a modulation axis oriented at -45° followed by a quarter wave retarder (QWR) oriented with fast axis at -90°. A fiber coupler and near-field fiber-probe are positioned before the sample, and a microscope objective (numerical aperture 0.85) is inserted just after the sample to collect the transmitted light. We

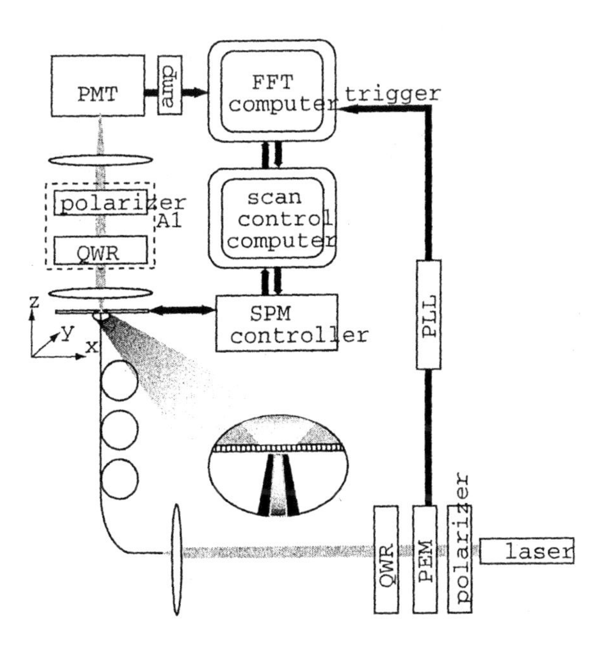

Figure 1. Schematic illustration of the NSOM polarimeter. A linear polarizer is followed by a PEM with modulation axis at -45° and quarter wave retarder (QWR) oriented with fast axis at -90°. A circular analyzer, A1, (QWR with fast axis at 0° and linear polarizer at -45°) follows the sample for birefringence measurements.. Fiber paddles are shown as loops. Inset illustrates the NSOM aperture probe.

use aluminum-coated pulled single-mode fiber probes (*40*) with aperture sizes from 50 nm to 180 nm. The aperture is held 5 nm to 10 nm from the sample using a shear-force feedback mechanism (*41*) that employs a small piezo-electric tube to sense tip motion. The sample is scanned using a piezo-driven flexure stage. To ensure a 1 MHz bandwidth (single pole), the detector, a photomultiplier tube (PMT), is run at high gain and the current amplifier at low gain (20 μA/V) and high bandwidth. A fast digital I/O board is used to acquire optical data; topographic data is acquired using a scanning probe microscope (SPM) control system (RHK SPM 1000 version 8).

At each pixel in an image, 8192 intensity data points (or more) spaced by $\pi/4\omega \cong 2.5$ μs are acquired. A phase-locked loop locks the sampling frequency to a multiple of the resonant PEM frequency, so the 1ω and 2ω components of a fast Fourier transform (FFT) of the intensity can be easily recovered. That is, the Fourier component representing the positive PEM frequency is commensurate with the 1024th point in our transform, and twice that frequency is commensurate with the 2048th point. As we will see below, these intensity modulations are the result of diattenuation or birefringence or both in the sample. An FFT of the intensity vs. time signal then yields the amplitude of the dc component and the amplitude and phase (or real and imaginary parts) of the 1ω and 2ω components, which are recorded. The digital I/O board and the SPM software are located on separate computers; hardware handshaking implemented between the two computers permits simultaneous acquisition of all optical and topographic data. Each point in the image takes 20 ms (for 8192 intensity points) to 80 ms (for 32k intensity points) to acquire so a 128 by 128 image takes a minimum of about 6 minutes to acquire; actual acquisition time (including deadtime during the handshake) was between 20 mins. and 2 hours.

Probe diattenuation and retardance must be considered when implementing polarization modulation in a near-field microscope. All near-field probes have diattenuation arising from asymmetries in the probe aperture or tip coating. An improperly coupled fiber will have additional diattenuation from reflections at the cleaved end (*31*), although careful cleaving and coupling can eliminate this. Fiber probes also have linear or circular retardance from strain birefringence or geometrical considerations in the fiber tail. The linear retardance of a non-diattenuating fiber can be nulled using a commercially available fiber polarization controller, sometimes called "fiber paddles" (e.g. Thorlabs Inc., FPC030). However, in the presence of a diattenuating tip, we show below that the fiber retardance cannot be nulled, only minimized; the residual birefringence must be accounted for. Circular birefringence of the fiber contributes an overall rotation of the incoming polarization, so absolute orientations cannot be determined a priori. In this work, *the diattenuating and fast axis orientations are always measured relative to an unknown but fixed axis (31).*

Polarization modulation polarimetry with Fourier analysis

In polarimetry, the change in polarization of light as it passes through a sample is measured and used to ascertain the properties of a material. A typical polarimeter places a sample between a polarization state generator and a polarization state analyzer. The polarimetric properties of a homogeneous, stationary material can be described by 8 parameters: a global phase change (corresponding to the average index-of-refraction of a material), a global absorption or transmission, two eigenpolarizations (polarizations for which light propagates through the materials with no polarization state change) specified by 4 parameters (each has an angle and a ellipticity), and the relative phase change and relative absorption (or relative transmittance) for eigenpolarized light (ratio of the eigenvalues). We concentrate here on 5 parameters that may be described by (1) the overall absorption or transmittance, (2) the relative absorption (dichroism) or transmittance (diattenuation) of the two eigenpolarizations, (3) linear birefringence (or more generally, the linear retardance), and (4) the orientations of the fast and (5) dichroic or diattenuating axes. The use of a polarization modulation technique (*42, 43*), described next, permits us to measure these 5 parameters in two passes over the sample (two different analyzer configurations).

For a linear diattenuator with transmittances given by q and r ($q>r$), we define diattenuation as:

$$D = \frac{q-r}{q+r}, \tag{1}$$

where q is the transmittance of light polarized parallel to the *diattenuating axis*, and r is the transmittance of light polarized perpendicular to this axis. Note, light polarized along the diattenuating axis has the higher transmittance. Likewise, for a linear retarder we define the retardance as a difference between phase shifts along the ordinary and extraordinary axis of retarder;

$$\theta = 2\pi \cdot (n_o - n') \cdot t / \lambda . \tag{2}$$

Here λ is the wavelength of light, t is the thickness of the sample, n_o is the ordinary index of refraction, and n' is the effective extraordinary index of the film. For a uniaxial crystal with the symmetry axis tilted at angle α from the propagation direction, n' is given by:

$$1/n' = \sqrt{\cos^2\alpha / n_o + \sin^2\alpha / n_e}, \tag{3}$$

where n_e is the extraordinary index (for light polarized along the symmetry axis). For a non-crystalline polymer we have

$$\theta = 2\pi \cdot \Delta n \cdot t \cdot f / \lambda, \tag{4}$$

where f is a factor characterizing the orientation of the chains; f = 1 for a perfectly aligned chain and f = 0 for random orientation. The intrinsic birefringence of the polymer, Δn, and the birefringence of the crystalline form (n_o-n_e) will generally be quite different.

Our instrument (described above, Fig. 1) is based on a polarization modulation polarimetry scheme described in Refs. (*42, 43*). The combination of linear polarizer, PEM and QWR results in light at the input to the fiber that is linearly polarized, with a polarization direction that is modulated at frequency

$\omega/2\pi$ Hz through an angle d given by the modulation amplitude of the PEM. This light passes through the tip and sample to be collected by the PMT. A circular analyzer, A1 in Fig. 1 (QWR oriented with its fast axis at 0° followed by a linear polarizer at -45°), may be inserted after the sample for measurements of retardance. If a linear retardance is present, the light will have a circular component whose amplitude changes with the input polarization direction. Accordingly, the detected signal after A1 will have harmonics of $\omega/2\pi$ Hz. To measure the linear diattenuation, A1 is removed and the intensity of the transmitted light at the detector is measured directly. An overall transmitted intensity that changes with incident linear polarization direction is an indicator of sample diattenuation and the magnitude and direction of the diattenuation can be extracted from the periodicity and amplitude of the signal. In general, the signal at the detector can be written

$$I(t) = I_0 + I_1 \sin(\omega t) + I_2 \cos(2\omega t) + \ldots \tag{5}$$

where I_0, I_1, and I_2 will be determined with and without A1 in place to determine the 5 parameters of interest to us here: the overall transmission, the diattenuation, the direction of the diattenuating axis, the linear retardance, and the direction of the "fast" or low index axis.

The parameters I_0, I_1, and I_2 are extracted from the FFT $[F(\nu)]$ of the intensity signal at the PMT;

$$F(0) = I_0 \tag{6}$$

$$\mathrm{Im}(F(\omega)) = -I_1/2 \tag{7}$$

$$\mathrm{Re}(F(2\omega)) = I_2/2 \tag{8}$$

With no sample and only the NSOM probe in place, we find the following relationships between the amplitudes I_0, I_1, and I_2 of Eqs. (5)–(8) and the retardance and diattenuation of the probe (*39*). With A1 removed:

$$R_{1\omega}^t \equiv I_1/I_0 = 2D_t J_1(d)[\sin(2\varphi_b^t)\cos(2\varphi_d^t - 2\varphi_b^t) + \cos(\theta_t)\cos(2\varphi_b^t)\sin(2\varphi_d^t - 2\varphi_b^t)] \tag{9}$$

$$R_{2\omega}^t \equiv I_2/I_0 = 2D_t J_2(d)[\cos(2\varphi_b^t)\cos(2\varphi_d^t - 2\varphi_b^t) - \cos(\theta_t)\sin(2\varphi_b^t)\sin(2\varphi_d^t - 2\varphi_b^t)] \tag{10}$$

Here $D_t = (u-v)/(u+v)$ is the diattenuation of the NSOM probe or tip and φ_d^t is the alignment angle of the diattenuating axis; θ_t is the retardance of the probe and φ_b^t is the alignment of the fast axis. The amplitude of the PEM phase modulation is given by d and the PEM frequency is $\omega/2\pi$ Hz. The amplitude $d = 2.405$ is chosen so that the zeroth order Bessel function $J_0(d) = 0$, which gives $J_1(d) \cong 0.519$ and $J_2(d) \cong 0.432$. Notice that both terms are proportional to D_t with factors that depend on both the orientation of the fast and diattenuating axes *and* the cosine of the probe retardance. As we demonstrate below, if the diattenuation of the tip is not too large, the retardance of the probe can be made small (but not zero), using the fiber-paddles and fiber-nulling procedure described by McDaniel *et al.* (*31*). To this end, we expand these equations for small θ_t (< 0.1) so that $\cos(\theta_t) \cong 1$. The error involved in this approximation will be of order θ_t^2, which for our probes will always be less than 0.01. In this case we can approximate Eqs. (9) and (10) by :

$$R_{1\omega}^t \cong 2D_t J_1(d)\sin(2\varphi_d^t) \tag{11}$$

$$R_{2\omega}^t \cong 2D_t J_2(d)\cos(2\varphi_d^t), \tag{12}$$

which we can use to determine the diattenuation of the tip.

For the case where A1 is in, we can show that the intensity at the PMT has Fourier components:

$$B_{1\omega}^t \equiv I_1 / I_0 = R_{1\omega}^t + 2J_1(d)\left(2\frac{\sqrt{uv}}{u+v}\sin(\theta_t)\cos(2\varphi_b^t)\right) \tag{13}$$

$$B_{2\omega}^t \equiv I_2 / I_0 = R_{2\omega}^t - 2J_2(d)\left(2\frac{\sqrt{uv}}{u+v}\sin(\theta_t)\sin(2\varphi_b^t)\right) \tag{14}$$

Here, it is apparent that most of dependence on diattenuation can be accounted for (without approximation) by a simple subtraction of the diattenuation measurement from the retardance measurement. This pattern repeats itself when measuring sample retardance in the presence of tip and sample diattenuation.

If the tip diattenuation is small, we can apply a further approximation to simplify the expressions above. For small diattenuation such that

$$\frac{\sqrt{uv}}{u+v} = \frac{1}{2}\sqrt{1-D_t^2} = \frac{1}{2}\left(1-\frac{1}{2}D_t^2+\ldots\right) \cong 1/2\,, \tag{15}$$

we see that Eqs. (13) and (14) can be written as:

$$B_{1\omega}^t \cong R_{1\omega}^t + 2J_1(d)\sin(\theta_t)\cos(2\varphi_b^t) \tag{16}$$

$$B_{2\omega}^t \cong R_{2\omega}^t - 2J_2(d)\sin(\theta_t)\sin(2\varphi_b^t)\,. \tag{17}$$

To minimize fiber probe birefringence and permit the use of the approximation that gives Eqs. (11) and (12), we adjust the fiber paddles shown to zero $B_{1\omega}^t$ and $B_{2\omega}^t$ within the noise limit ("nulling"). If $D_t = 0$, this would guarantee $\theta_t = 0$. D_t is never zero but for most tips it is below 0.1. Therefore nulling gives $R_{1\omega}^t \cong -2J_1(d)\left(\sin(\theta_t)\cos(2\varphi_b^t)\right)$ and $R_{2\omega}^t \cong 2J_2(d)\left(\sin(\theta_t)\sin(2\varphi_b^t)\right)$. Using Eqs. (11) and (12) and recalling that $d = 2.405$, we find that that $|\sin(\theta_t)| \approx D_t$. Eqs. (11) and (12) can therefore be used after nulling if $D_t < 0.1$.

To measure the retardance and diattenuation of a sample, we assume: (1) that samples are both diattenuating and birefringent, but that the retardance and diattenuation are small – a reasonable expectation for thin polymer films and (2) that $D_t < 0.1$, so that θ_t can be kept small as discussed above. In this case, the full expression for the ratios of the Fourier components of the measured intensity are, for A1 removed (diattenuation measurement) (*39*):

$$R_{1\omega}^s \cong 2J_1(d)\left[D_t\sin(2\varphi_d^t) + D_s\sin(2\varphi_d^s)\right]; \tag{18}$$

$$R_{2\omega}^s \cong 2J_2(d)\left[D_t\cos(2\varphi_d^t) + D_s\cos(2\varphi_d^s)\right]. \tag{19}$$

Here D_s is the sample diattenuation and φ_d^t is the orientation of the diattenuating axis. We use the same approximations as above for both tip and sample diattenuation and retardance, which gives Eqs. (18) and (19) correct to second order in θ_s, θ_t, D_t, D_s, and their products. Similarly, the ratios of the Fourier components for a retardance measurement (A1 in) are as follows:

$$B^s_{1\omega} \cong R^s_{1\omega} + 2J_1(d)\left[\sin(\theta_t)\cos(2\varphi^t_b) + \sin(\theta_s)\cos(2\varphi^s_b)\right]; \tag{20}$$

$$B^s_{2\omega} \cong R^s_{2\omega} - 2J_2(d)\left[\sin(\theta_t)\sin(2\varphi^t_b) + \sin(\theta_s)\sin(2\varphi^s_b)\right]. \tag{21}$$

Where θ_s is the sample retardance and φ^s_b the orientation of the sample fast axis. Eqs. (18) and (19) (and the tip properties) are used to arrive at the sample diattenuation. To determine the sample retardance, we first measure the diattenuation of the sample and then subtract $R^s_{1\omega}$ and $R^s_{2\omega}$ directly from the result of our retardance measurement, as suggested by Eqs. (20) and (21). Note that the tip diattenuation need not be explicitly accounted for in the retardance measurement if this procedure is followed. A more complete discussion of this analysis can be found in Ref. (*39*).

NFP of Photonic Block Copolymer Morphology

A study of block copolymer (BC) lamellar morphology by NFP (*32*) is summarized here. Microphase separation in BCs, driven by the immiscibility of the end-connected constituent polymer chains or *blocks*, produces a variety of domain motifs (lamellae, double-gyroid, hexagonal-packed cylinders and BCC spheres) with a related set of 1-, 2- and 3-dimensional band structures (*44-46*), tunable through the BC composition (*35, 47*). A BC's molecular mass (M_r) governs the microphase domain periodicity (L_0), typically limited to a range of 10 nm - 100 nm. However, recent synthetic efforts have produced ultra-high molecular mass BCs with L_0 of 150 nm - 300 nm, enabling M_r-tailored photonic band gaps in the visible (*33, 48, 49*). This morphological flexibility is complimented by an extensive set of techniques geared to perfect/control the order of BC structures and furnish them with added functionality (*50*). These strategies can be harnessed to enhance the optical performance of BC materials and devices based upon them. The optical activity of single microphase domains and defect structures may dictate device function; the advance of photonic BC systems requires a technique to characterize optical properties at the mesoscale.

Polystyrene-b-polyisoprene (PS-b-PI) block copolymer ($M_r = 1.4\times10^6$) used here is nearly volume-symmetric in composition (PS/PI = 480K/560K), and exhibits the lamellar microdomain motif with $L_0 \approx 240$ nm. Bulk specimens were processed by roll casting, which helps order and align the domains (*51*). Thin (100 nm) sections were sliced from the bulk using cryo-ultramicrotomy and deposited onto glass coverslips. Subsequent exposure to OsO_4 vapor (2 hours) preferentially crosslinks the PI domains, making them less mechanically compliant (more amenable to shear-force feedback), and enhancing the optical contrast between PS and PI. Single lamellar domains and defects are resolved in transmission (Fig. 2b). The PI domains ($n = 1.52$, stained) appear darker, while PS domains ($n = 1.59$) appear lighter, as verified through the topography image (Fig. 2d) and plot in Fig. 2, which demonstrate that darker domains are also lower in height, due to PI contraction during OsO_4 crosslinking.

Fig. 2a maps the diattenuation of a BC specimen (analyzed using Eqs. 18 and 19) with simultaneously acquired topography and transmission micrographs. The optical images (Fig. 2a-c), and in particular the diattenuation image (Fig. 2a), provide excellent morphological detail. Comparison of the diattenuation (Fig. 2a) and transmission (Fig. 2b) along the white line (Fig. 2, right) show that the higher-transmitting PS domains appear most diattenuating (3 % to 5 %), the absorbing PI domains appear less diattenuating (2 % to 4 %), and that D_s is minimized near the domain interface. An explanation for this pattern of D_s is suggested by the Bethe-Bowkamp model (BB) (*52, 53*), which approximates the field at an NSOM aperture. In BB, the field pattern at the tip is elongated along the polarization axis. This anisotropic field pattern centered over an absorbing PI domain, transmits less/more when polarized parallel/perpendicular to the domain, producing apparent diattenuation (*39*). For the less absorbing PS domains, the opposite is true. Indeed, φ_d^s (Fig. 2c) alternates between domains with a difference $\approx 90°$. The diattenuation here is therefore not intrinsic to the sample but an artifact of the tip/sample interaction.

Fig. 3 focuses on a symmetric tilt boundary where the lamella bend through two "kinks" along an "N"-shaped track. The diattenuation images (c,e) are used (with probe data) to correct the birefringence images (d,f), via Eqs. (20) and (21). As in Fig. 2, the D_s image (Fig. 3c) illuminates the domain and interface morphology across the defect. The retardance (θ_s) is mapped in Fig. 3d. The standard uncertainty is ± 10 mrad (*39*). Contrast here is governed by Eq. (4) which relates θ_s to the sample thickness (t), the intrinsic birefringence (Δn), the illumination wavelength ($\lambda = 488$ nm) and the degree of chain orientation (f) (*54*). Inter-domain contrast is therefore based upon the difference in Δn between PS ($\Delta n = 0.195$ for atactic PS) (*54*) and PI ($\Delta n = 0.13$) (*55*), and f, which reflects the net average elongation of chains perpendicular to the interface exhibited by ordered BC systems. Thus, PS domains appear lighter (larger θ_s), while PI domains appear darker, as verified by comparing Fig. 3d with Fig. 3b. Contrast across the defect is dictated by f, and t. In principle θ_s is proportional to the local stress (through f), but variations in the through-plane lamellar orientation, *i.e.* "projection effects" due to arbitrary sectioning of the specimen, may also affect θ_s. Indeed, as expected with projection effects, θ_s often seems inversely related

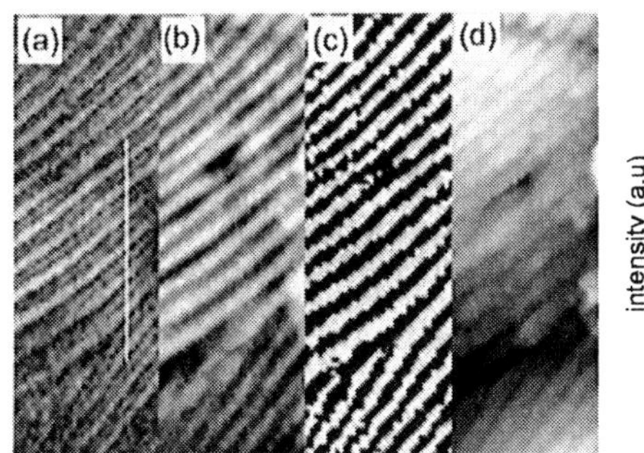

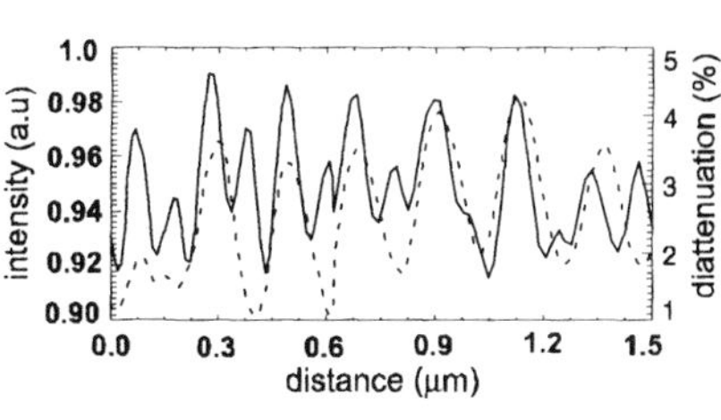

Figure 2. Left: NFP data from a BC specimen. (a) diattenuation [0 % -5 %], (b) transmission (normalized intensity) [0.6-1.0], (c) orientation of the diattenuating axis orientation [0°-180°], (d) topography [0 nm – 25 nm]. Right: diattenuation (solid line) and transmitted intensity (dotted line) along the white line shown in (a). (See page 1 of color insert.)

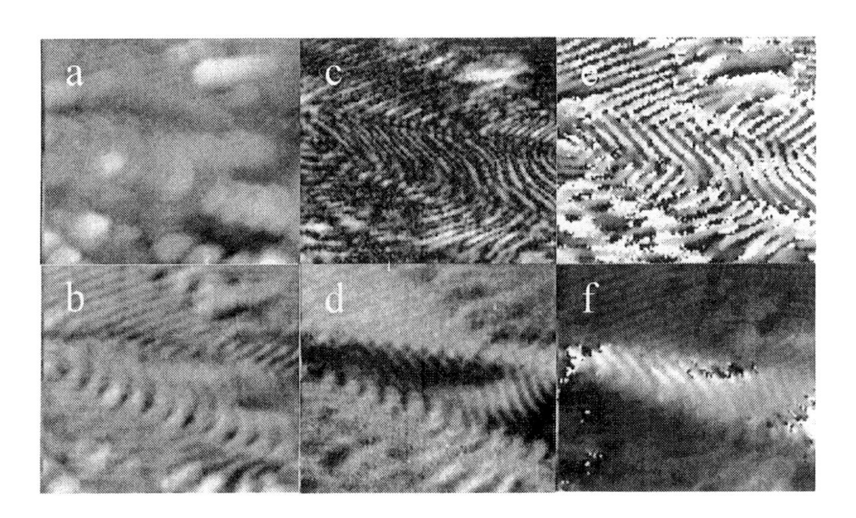

Figure 3. NFP images of BC sample showing a symmetric tilt boundary. (a) topography [0 nm – 25 nm], (b) transmission [0.8 – 1.0], (c) diattenuation [0 % - 9 %], (d) retardance [0 mrad – 122 mrad], (e) relative angle of diattenuating axis [0°-180°], (f) relative angle of fast axis [0°-180°]. Scan size is 3.0 μm by 3.6 μm. (See page 1 of color insert.)

to the apparent L_0 observed in this image. In-plane lamellar orientation is illuminated in Fig. 3f, which maps the relative φ_b^s. Due to chain elongation, the low-n axis lies perpendicular to the domain interface; φ_b^s reflects this. As the lamellae bend though the symmetric tilt boundary, φ_b^s increases and decreases accordingly. Discontinuities in the φ_b^s image are due to phase wrapping.

NFP of polymer crystallites

The characteristic spherulite crystallization pattern of bulk polymers has been studied for many years, but other morphologies, especially those that occur in otherwise amorphous thin films undergoing "cold" crystallization, present measurement difficulties (*56-62*) and so have been less studied. Traditional methods most often lack the spatial resolution and/or sensitivity required to examine 1) chain conformations near the growth front, 2) amorphous layers posited to exist between lamellae, and 3) the orientation of folded chains within these crystallites. NFP is well suited to the study of thin films and can be used to elucidate structure and the character of the strain field in these non-equilibrium crystallization patterns.

Polymers crystallize by forming folded layers (lamellae). In isotactic polystyrene (iPS), studied here and in Refs. (*63, 64*), these lamellae organize on a larger scale into hexagonal crystals with 6-fold symmetry around the chain (c) axis. Polystyrene is therefore a uniaxial crystal with a fast c axis. In a sufficiently thin film (thickness less than the radius of gyration of the polymer, R_g), crystallites one lamellar thickness will form in a variety of hexagonally symmetric and branching morphologies, generally with the c axis perpendicular to the substrate. At the top and bottom surfaces of each lamella, the polymer chains loop back on themselves and a thin amorphous layer should be expected and has been reported (*65*). A boundary depleted of polymer (depletion boundary) forms around the growing crystallite as polymer from the less-dense amorphous region is pulled into the crystallite (*65*). In thicker films spherulites can form and the lamella stack and the chain axis can tilt or twist (*66*). An amorphous layer above the folded lamellae and depletion region around the growing crystal pattern should still be present.

Two separate morphologies, an early-growth spherulite and a dendritic, "compact seaweed" morphology (*67-69*), are shown here in Figs. 4-6. Samples were prepared as discussed in Refs. (*63, 64*). Film thickness was approximately 85 nm for the spherulites and 15 nm for the dendrites.

Fig. 4a shows shear-force images of typical crystallites with seaweed morphology (*67-69*) taken with a sharp NSOM probe (*40*). The thickness of

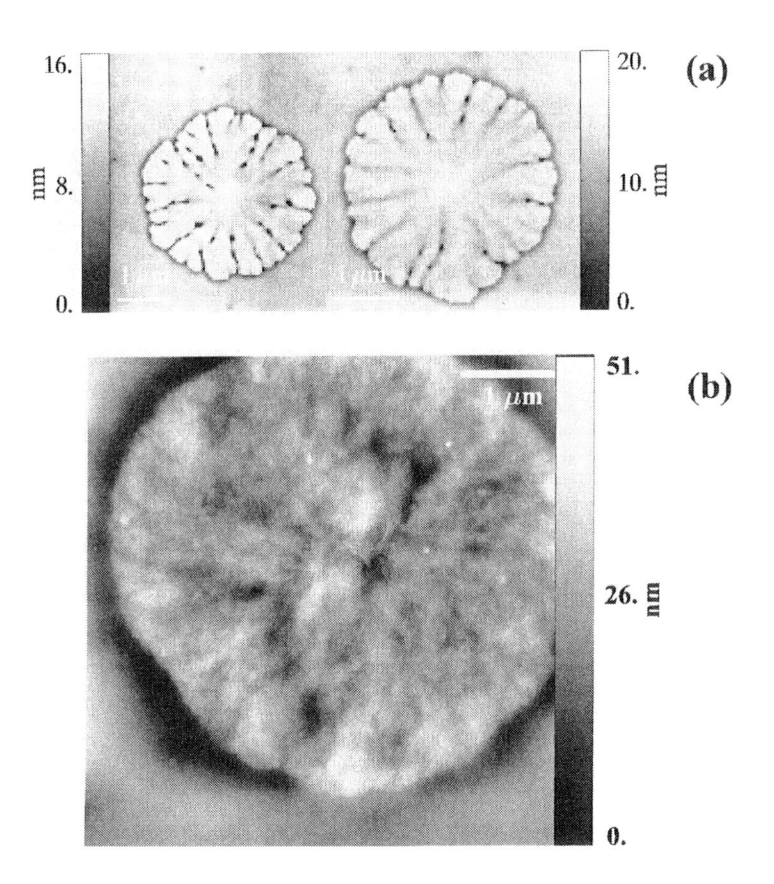

Figure 4. Topographic images of polymer crystallites studied here. Top: dendritic crystallites with compact seaweed morphology, images acquired using shear-force microscopy and an NSOM tip. Bottom: early-growth stage spherulite. Image acquired using AFM. (See page 2 of color insert.)

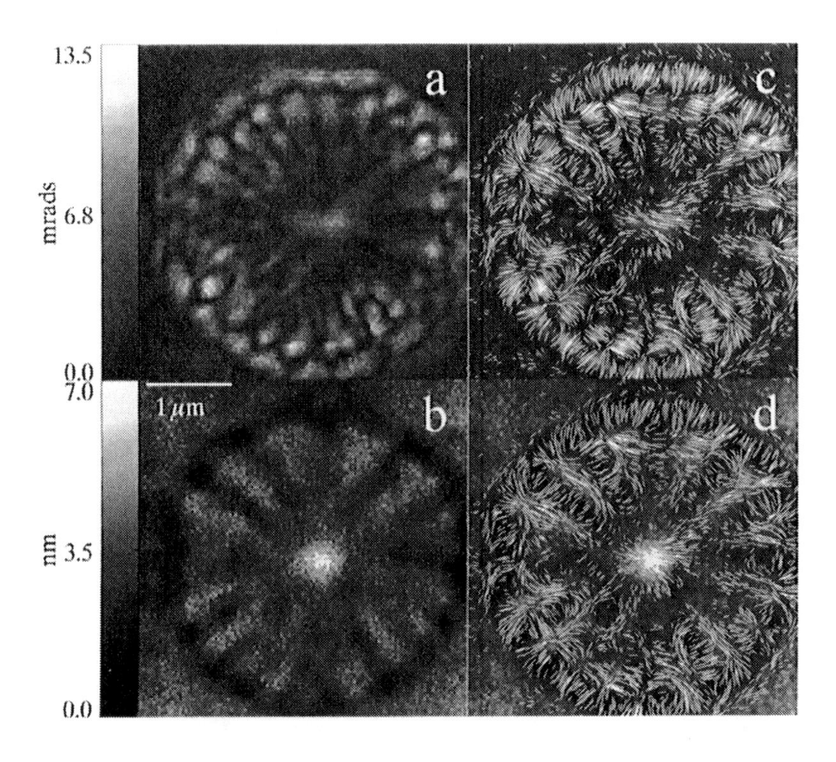

Figure 5 (Left). NFP images of compact seaweed morphology dentrite. (a) retardance, (b) topography, (c) retardance with overlaid fast axis orientation marks, (d) topography with overlaid fast axis orientation marks. Fast axis alignment shown for $\theta > 2.0$ mrad. (See page 3 of color insert.)

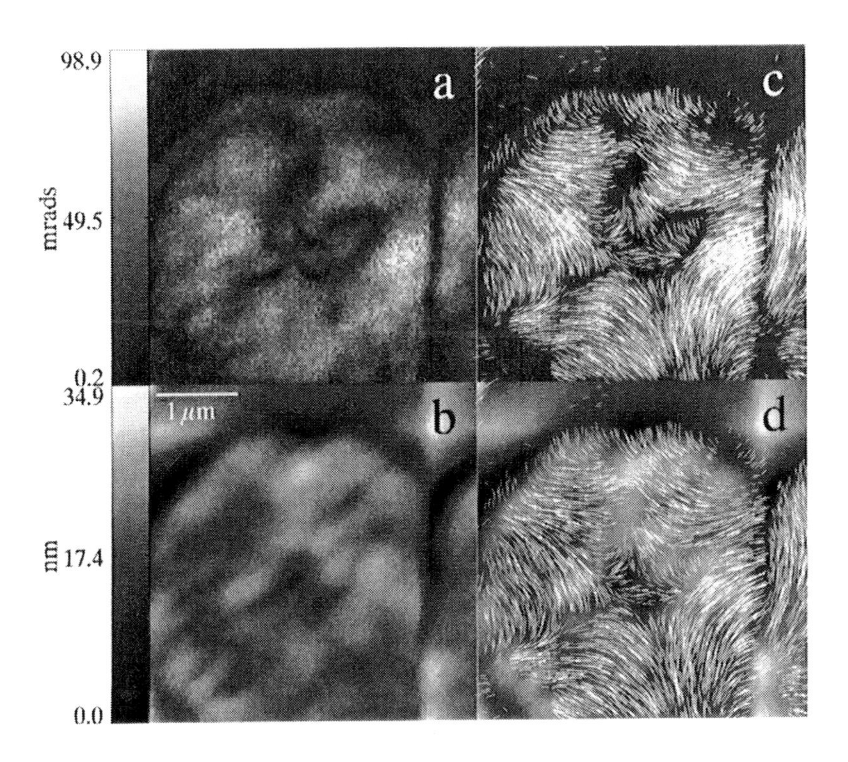

Figure 6. NFP images of an early-growth spherulite. (a) retardance, (b) topography, (c) retardance with overlaid fast axis orientation marks, (d) topography with overlaid fast axis orientation marks. Fast axis alignment shown for θ > 20 mrad. (See page 4 of color insert.)

these films is less than the radius of gyration for the iPS (R_g = 22 nm), so the crystallites are a single lamella thick. Fig. 4b shows an AFM image of a typical spherulite. In both cases, depletion regions, with low topography, are seen around the edges of the pattern and a thick nucleation site is visible in the center.

NFP images of a seaweed pattern are shown in Fig. 5. The topography image (Fig. 5b) is lower resolution that those of Fig. 1; good NSOM tips are flat with a diameter of approximately 400 nm (*40*), giving poor topographic images. Fig. 5a shows a birefringent structure in an otherwise amorphous ($\theta = 0$) film. Birefringence is highest at the center of pattern (the nucleation site), around the edges (in the depletion boundary, where 6 mrad < θ < 10 mrad) and near the growth tips. The noise floor (standard deviation) is 0.6 mrad for this image. The fast axis orientation is shown with lines overlaid on the retardance and topography images in Figs. 5c and d. The fast axis lines are drawn with length proportional to θ, and have been omitted where the θ < 2.0 mrad. An overall shift of 0.45 rad has been applied to the fast axis orientation to bring polymer in the depletion region into radial alignment.

In this image the main source of birefringence is most likely strain in the amorphous layers. In similar crystallites, Taguchi *et al.* have found that the c axis is always within 6° of normal to the surface (*70-72*), which would result in θ < 1 mrad from Eq. 2 and 3 [for iPS, $\langle n \rangle$ = 1.6 (*54*) and n_o-n_e = 0.28 (*73, 74*); the intrinsic birefringence Δn = 0.167 (*54, 73, 75*)]. A stressed amorphous layer only 5 nm thick has maximum θ = 10 mrad (for f = 1, from Eq. 4); a 15 nm thick strained amorphous film has maximum θ = 32 mrad. The stress in amorphous regions or layers can account for most if not all of the retardance in Fig. 5.

In Figure 6 we show NFP images of a spherulite. A second crystallite is adjacent on the right side. If a –1.12 rad shift is applied globally to the fast axis orientation data, we again see a depletion region consistent with radial strain. The retardance in this region is 20 mrad < θ < 30 mrad. Here the uncertainty on the retardance (one standard deviation) is 5 mrad. Healing of the fast axis direction from radial to circumferential as we move in radially towards the center of the spherulite is evident, except at 5 angles at which the radial alignment persists to the center. For two other (isolated) spherulites, there were 6 such incursions (*63*). For amorphous iPS with f = 1, the maximum retardance for an 84 nm thick film is 183 mrad (Eq. 4), roughly twice the measured maximum. The retardance measured in Fig. 6 is greater than could possibly be due to strain in a thin amorphous layer but less than is possible for iPS perfectly aligned with c axis parallel to the substrate. It seems likely that θ reflects a combination of strain in the amorphous regions and tilt of the chain axis in the crystalline form.

Summary and Conclusions

In many respects, NFP is a natural extension of classical micro-polarimetry that has been used with great success to understand stress and crystallization in thicker polymer materials (*66*). We expect other innovations first pioneered using far-field techniques might be adapted for near-field use and correspondingly greater resolution. The adaptation of Fourier polarimetry, using polarization modulation and a real-time FFT, already has several advantages over static techniques or the use of lock-in amplifiers. First, it is a more flexible system, permitting extension of the Fourier analysis to more frequency components, and therefore more optical properties, without adding electronics. Second, this setup permits the use of a single input channel to collect all the polarimetric data. This concern is particularly relevant to the integration of polarimetry with NSOM, where each input channel must be synchronized with the position of a scanning stage and the number of input channels is often limited. This design makes it possible to easily incorporate more generalized polarimetry, such as that described in Refs. (*76, 77*) since an arbitrary number of Fourier components can be monitored without need for further input channels.

We have shown how NFP can be used to map out the linear diattentuation and birefringence of thin polymer samples, and we have briefly discussed the origins of these optical properties. Quantitative modeling that takes into account both the physical properties of the films and artifacts due to tip-sample interactions remain to be developed.

SNG acknowledges a NRC/NIST Post-doctoral Research Fellowship. Funding was through the NIST Advanced Technology Program and Physics Laboratory. The authors thank K.L. Beers of the Polymers Division, NIST, for assistance in preparing the film specimens, J.F. Douglas, of the same division, and G.W. Bryant, of the Atomic Physics Division, NIST, for illuminating discussions and physical insight.

References

1. E. Betzig, A. Lewis, A. Harootunian, et al. *Biophys. J.* **1986**, *49*, 269-279.
2. D. W. Pohl, W. Denk and M. Lanz. *Appl. Phys. Lett.* **1984**, *44*, 651-653.

3. E. Betzig, J. K. Trautman, J. S. Weiner, et al. *Appl. Optics* **1992**, *31*, 4563-4568.
4. M. Vaeziravani and R. Toledocrow. *Appl. Phys. Lett.* **1993**, *63*, 138-140.
5. R. Toledocrow, J. K. Rogers, F. Seiferth, et al. *Ultramicroscopy* **1995**, *57*, 293-297.
6. G. A. Valaskovic, M. Holton and G. H. Morrison. *J. Microsc.-Oxf.* **1995**, *179*, 29-54.
7. T. J. Silva and S. Schultz. *Rev. Sci. Instrum.* **1996**, *67*, 715-725.
8. E. Betzig, J. K. Trautman, R. Wolfe, et al. Appl. Phys. Lett. 1992, 61, 142-144.
9. A. Jalocha and N. F. van Hulst. *J. Opt. Soc. Am. B-Opt. Phys.* **1995**, *12*, 1577-1580.
10. D. A. Higgins, P. J. Reid and P. F. Barbara. *J. Phys. Chem.* **1996**, *100*, 1174-1180.
11. J. A. DeAro, K. D. Weston, S. K. Buratto, et al. *Chem. Phys. Lett.* **1997**, *277*, 532-538.
12. J. A. Teetsov and D. A. Vanden Bout. *J. Am. Chem. Soc.* **2001**, *123*, 3605-3606.
13. J. Teetsov and D. A. Vanden Bout. *Macromol. Symp.* **2001**, *167*, 153-166.
14. J. Teetsov and D. A. Vanden Bout. *Langmuir* **2002**, *18*, 897-903.
15. D. A. Higgins, X. M. Liao, J. E. Hall, et al. *J. Phys. Chem. B* **2001**, *105*, 5874-5882.
16. E. Mei and D. A. Higgins. *J. Chem. Phys.* **2000**, *112*, 7839-7847.
17. H. Ade, R. ToledoCrow, M. VaezIravani, et al. *Langmuir* **1996**, *12*, 231-234.
18. D. A. Higgins, D. A. VandenBout, J. Kerimo, et al. *J. Phys. Chem.* **1996**, *100*, 13794-13803.
19. T. Lacoste, T. Huser, R. Prioli, et al. *Ultramicroscopy* **1998**, *71*, 333-340.
20. T. Lacoste, T. Huser and H. Heinzelmann. *Z. Phys. B-Condens. Mat.* **1997**, *104*, 183-184.
21. T. Huser, T. Lacoste, H. Heinzelmann, et al. *J. Chem. Phys.* **1998**, *108*, 7876-7880.
22. P. K. Wei and W. S. Fann. *J. Microsc.-Oxf.* **2001**, *202*, 148-153.
23. C. H. Tan, A. R. Inigo, J. H. Hsu, et al. *J. Phys. Chem. Solids* **2001**, *62*, 1643-1654.
24. P. K. Wei, Y. F. Lin, W. Fann, et al. *Phys. Rev. B* **2001**, *6304*, 045417.
25. T. J. Silva, S. Schultz and D. Weller. *Appl. Phys. Lett.* **1994**, *65*, 658-660.
26. V. Kottler, N. Essaidi, N. Ronarch, et al. *J. Magn. Magn. Mater.* **1997**, *165*, 398-400.
27. P. Fumagalli, A. Rosenberger, G. Eggers, et al. *Appl. Phys. Lett.* **1998**, *72*, 2803-2805.

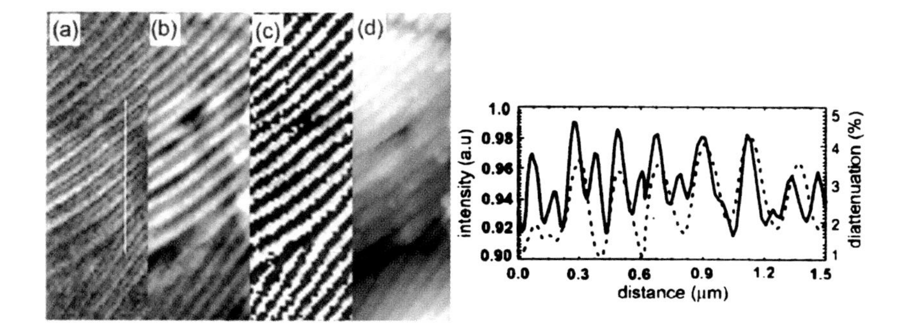

Figure 6.2. Left: NFP data from a BC specimen. (a) diattenuation [0 % -5 %], (b) transmission (normalized intensity) [0.6-1.0], (c) orientation of the diattenuating axis orientation [0°-180°], (d) topography [0 nm – 25 nm]. Right: diattenuation (solid line) and transmitted intensity (dotted line) along the white line shown in (a).

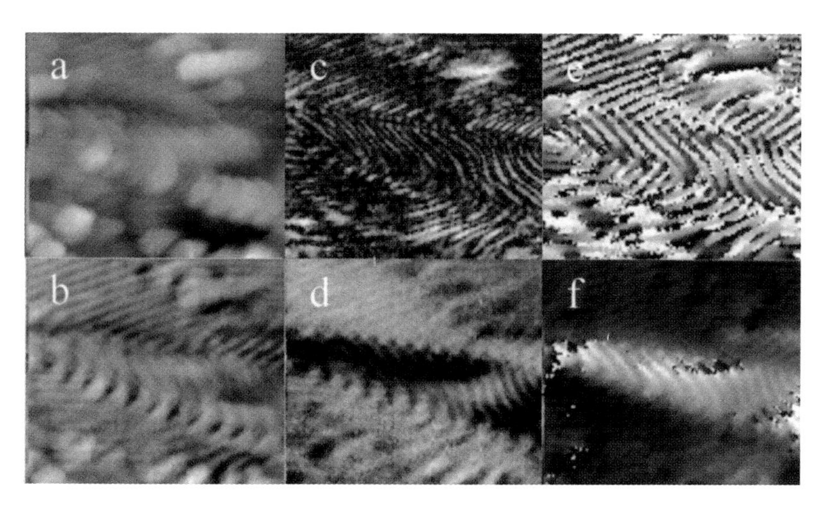

Figure 6.3. NFP images of BC sample showing a symmetric tilt boundary. (a) topography [0 nm – 25 nm], (b) transmission [0.8 – 1.0], (c) diattenuation [0 % - 9 %], (d) retardance [0 mrad – 122 mrad], (e) relative angle of diattenuating axis [0°-180°], (f) relative angle of fast axis [0°-180°]. Scan size is 3.0 μm by 3.6 μm.

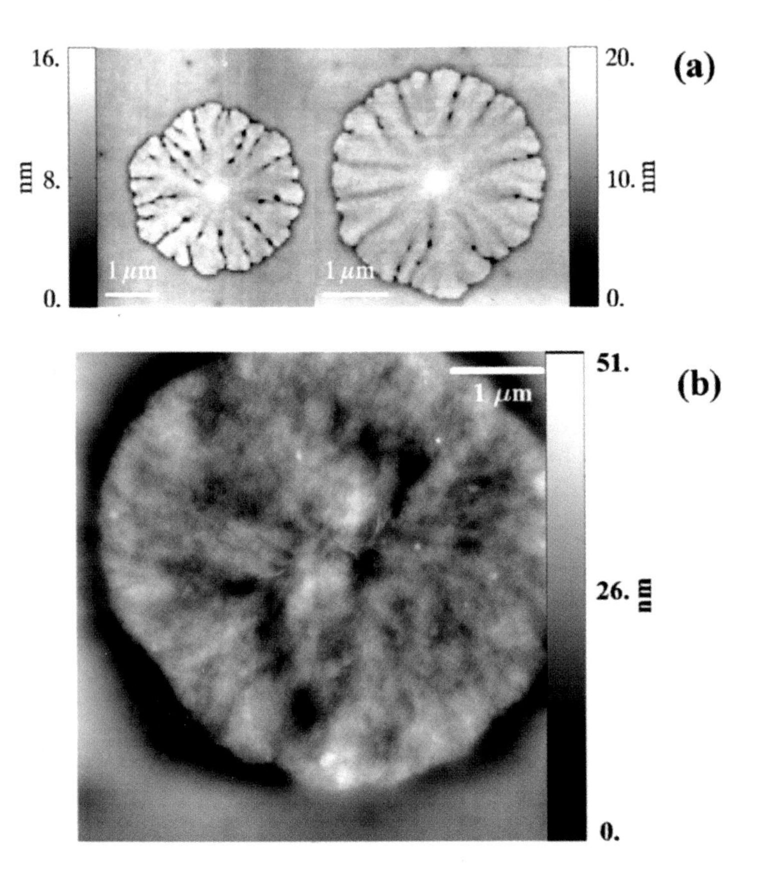

Figure 6.4. Topographic images of polymer crystallites studied here. Top: dendritic crystallites with compact seaweed morphology, images acquired using shear-force microscopy and an NSOM tip. Bottom: early-growth stage spherulite. Image acquired using AFM.

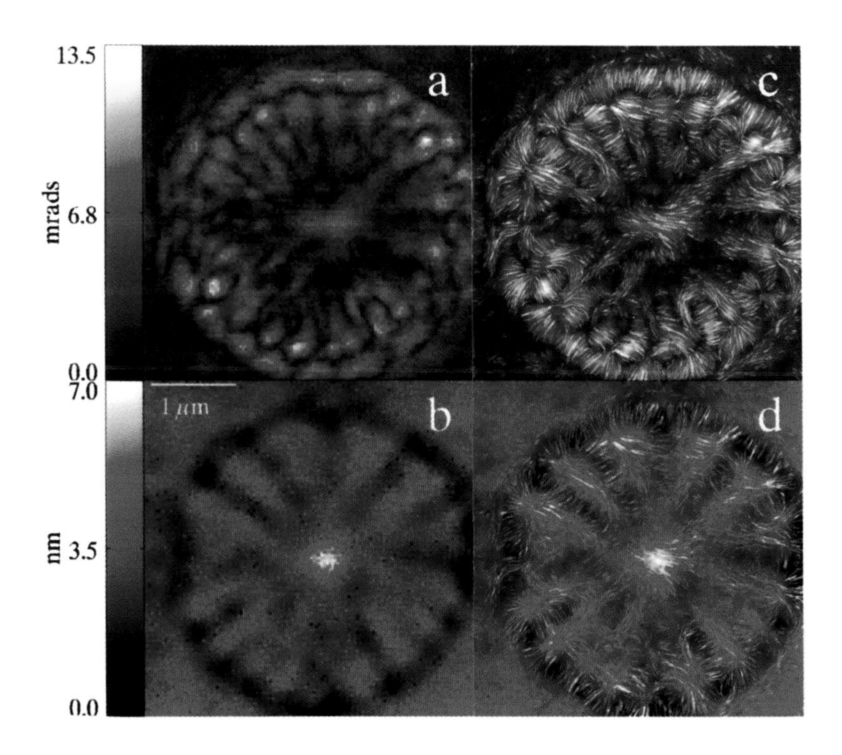

Figure 6.5 (Left). NFP images of compact seaweed morphology dentrite. (a) retardance, (b) topography, (c) retardance with overlaid fast axis orientation marks, (d) topography with overlaid fast axis orientation marks. Fast axis alignment shown for θ > 2.0 mrad.

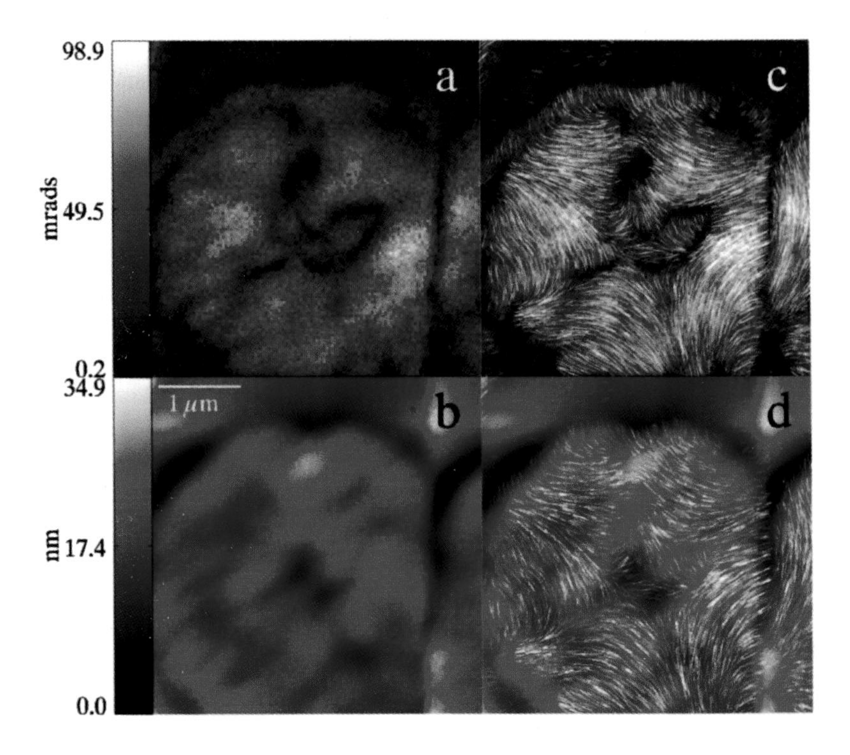

Figure 6.6. NFP images of an early-growth spherulite. (a) retardance, (b) topography, (c) retardance with overlaid fast axis orientation marks, (d) topography with overlaid fast axis orientation marks. Fast axis alignment shown for θ > 20 mrad.

28. O. Bergossi, H. Wioland, S. Hudlet, et al. *Jpn. J. Appl. Phys. Part 2 - Lett.* **1999**, *38*, L655-L658.
29. T. Roder, L. Paelke, N. Held, et al. *Rev. Sci. Instrum.* **2000**, *71*, 2759-2764.
30. L. Ramoino, M. Labardi, N. Maghelli, et al. *Rev. Sci. Instrum.* **2002**, *73*, 2051-2056.
31. E. B. McDaniel, S. C. McClain and J. W. P. Hsu. *Appl. Optics* **1998**, *37*, 84-92.
32. M. J. Fasolka, L. S. Goldner, J. Hwang, et al. *Phys. Rev. Lett.* **2003**, *90,* 016107.
33. A. Urbas, Y. Fink and E. L. Thomas. *Macromolecules* **1999**, *32*, 4748-4750.
34. A. C. Edrington, A. M. Urbas, P. DeRege, et al. *Adv. Mater.* **2001**, *13*, 421-425.
35. F. S. Bates and G. H. Fredrickson. *Annu. Rev. Phys. Chem.* **1990**, *41*, 525.
36. D. W. Pohl. *Advances in optical and electron microscopy* **1991**, *12*, 243.
37. E. Betzig and J. K. Trautman. *Science* **1992**, *257*, 189-195.
38. R. C. Dunn. *Chem. Rev.* **1999**, *99*, 2891-+.
39. L. S. Goldner, M. J. Fasolka, S. Nougier, et al. *Appl. Optics* **2003**, *42*, 3864-3881.
40. E. Betzig, J. K. Trautman, T. D. Harris, et al. *Science* **1991**, *251*, 1468-1470.
41. E. Betzig, P. L. Finn and J. S. Weiner. *Appl. Phys. Lett.* **1992**, *60*, 2484-2486.
42. P. L. Frattini and G. G. Fuller. *J. Colloid Interface Sci.* **1984**, *100*, 506-518.
43. S. J. Johnson, P. L. Frattini and G. G. Fuller. *J. Colloid Interface Sci.* **1985**, *104*, 440-455.
44. Y. Fink, A. M. Urbas, M. G. Bawendi, et al. *J. Lightwave Technol.* **1999**, *17*, 1963-1969.
45. M. Maldovan, A. M. Urbas, M. Yufa, et al. *Phys. Rev. B* **2002**, *65*, 165123.
46. A. M. Urbas and E. L. Thomas. Bicontinuous Cubic Photonic Crystal in a Block Copolymer System, 2002
47. E. L. Thomas and R. L. Lescanec. *Philos. Trans. R. Soc. Lond. Ser. A-Math. Phys. Eng. Sci.* **1994**, *348*, 149-166.
48. A. Urbas, R. Sharp, Y. Fink, et al. *Adv. Mater.* **2000**, *12*, 812-814.
49. A. C. Edrington, A. M. Urbas, P. DeRege, et al. *Adv. Mater.* **2001**, *13*, 421.
50. M. J. Fasolka and A. M. Mayes. *Annu. Rev. Mat. Res.* **2001**, *31*, 323.
51. R. J. Albalak and E. L. Thomas. *J. Polym. Sci. Pt. B-Polym. Phys.* **1994**, *32*, 341-350.
52. H. A. Bethe. *Phys. Rev.* **1944**, *66*, 163.
53. C. J. Bowkamp. *Philips Res. Rep.* **1950**, *5*, 321.
54. *Polymer Handbook*; 4th ed.; J. Brandrup, E. H. Immergut and E. A. Grulke, Eds.; Johh Wiley & Sons, Inc.: New York, 1999

55. M. Hashiyama, R. G. Gaylord and R. S. Stein. *Makromol. Chem. Suppl.* **1975**, *1*, 579.
56. K. Izumi, G. Ping, M. Hashimoto, et al. Crystal Growth of Polymers in Thin Films. In *Advances in Understanding of Crystal Growth Mechanisms*; T. Nishinaga, K. Nishioka, J. Harada, A. Sasaki and H. Takei, Eds.; Elsevier Science: Amsterdam, 1997; pp 337-348.
57. R. L. Jones, S. K. Kumar, D. L. Ho, et al. *Macromolecules* **2001**, *34*, 559-567.
58. O. Mellbring, S. K. Oiseth, A. Krozer, et al. *Macromolecules* **2001**, *34*, 7496.
59. G. Reiter. *Europhys. Lett.* **1993**, *23*, 579-584.
60. G. Reiter and J.-U. Sommer. *J. Chem. Phys.* **2000**, *112*, 4376-4383.
61. Y. Sakai, M. Imai, K. Kaji, et al. *J. Cryst. Growth* **1999**, *203*, 244.
62. S. Sawamura, H. Miyaji, K. Izumi, et al. *J. Phys. Soc. Japan* **1998**, *67*, 3338-3344.
63. L. S. Goldner, S. N. Goldie, M. J. Fasolka, et al. *in press, Appl. Phys. Lett.* **2004**.
64. S. N. Goldie, M. J. Fasolka, L. S. Goldner, et al. *Polym. Mater. Sci. Eng.* **2003**, *88*, 145.
65. K. Izumi, P. Gan, A. Toda, et al. *Japanese Journal of Applied Physics* **1994**, *33*, L 1628 - L 1630.
66. B. Wunderlich *Macromolecular Physics*; Academic Press: New York, 1973
67. E. Brener, H. Muller-Krumbhaar, D. Tempkin, et al. *Physica A* **1998**, *249*, 73-81.
68. V. Ferreiro, J. F. Douglas, J. Warren, et al. *Phys. Rev. E* **2002**, *65*.
69. V. Ferreiro, J. F. Douglas, J. A. Warren, et al. *Phys. Rev. E* **2002**, *65*.
70. K. Taguchi, H. Miyaji, K. Izumi, et al. *Polymer* **2001**, *42*, 7443-7447.
71. K. Taguchi, H. Miyaji, K. Izumi, et al. *J. Macromol. Sci.-Phys.* **2002**, *B41*, 1033-1042.
72. K. Taguchi, Y. Miyamoto, H. Miyaji, et al. *Macromolecules* **2003**, *36*, 5208-5213.
73. E. F. Gurnee. *J. Appl. Phys.* **1954**, *25*, 1232-1240.
74. R. S. Stein. *J. Appl. Phys.* **1961**, *32*, 1280-1286.
75. V. N. Tsevtkov. *Journal of Polymer Science* **1962**, *57*, 727-741.
76. G. E. Jellison and F. A. Modine. *Appl. Optics* **1997**, *36*, 8184-8189.
77. G. E. Jellison and F. A. Modine. *Appl. Optics* **1997**, *36*, 8190-8198.

Mechanical Studies of Polymers by Atomic Force Microscopy

Chapter 7

Dynamic Atomic Force Microscopy Analysis of Polymer Materials: Beyond Imaging Their Surface Morphology

Ph. Leclère[1,2], V. Cornet[1], M. Surin[1], P. Viville[1], J. P. Aimé[2], and R. Lazzaroni[1]

[1]Service de Chimie des Matèriaux Nouveaux, Université de Mons-Hainaut. Materia Nova, Place du Parc 20, B–7000 Mons, Belgium
[2]CPMOH, Unversité de Bordeaux I, 351 Cours de la Libération, F–33405 Talence Cedex, France

Dynamic atomic force microscopy is known for its ability to image soft materials without inducing severe damage. The understanding of the origin of the image contrast is not obvious and constitutes an important subject of debate. Here, we propose a straightforward method, based on the analysis of approach-retract curves, which provides an unambiguous quantitative measurement of the local mechanical response and/or topographic contribution(s), depending on the studied sample. From the recorded data, we show that it is possible to determine the different contributions and, therefore, go beyond the morphological aspects. This approach is illustrated here on a thermoplastic elastomer block copolymer, used as a model system presenting phase-separated nanodomains characterized by specific mechanical properties. The extension of the technique to other polymer systems, such as polymer blends, polymer nanocomposites, and conjugated materials, is also discussed.

Introduction

The development of nanotechnology implies a large effort to study and understand physical phenomena at the nanometer scale. Methods using local force probes provide important contributions to those studies, because the small size of the tip allows one to probe surfaces with excellent lateral and vertical resolution. Among those scanning microscopies, dynamical force techniques, *i.e.*, using an oscillating tip, are particularly well adapted to soft samples such as polymers or biological systems. Among many applications of those techniques, one convincing illustration of their potential is the advance brought to microstructural studies of block copolymers by phase imaging in Tapping mode Atomic Force Microscopy [1-7].

In dynamical modes, two types of operation are possible: either the oscillating amplitude is fixed and the output signal is the resonance frequency (this is called the non-contact resonant force mode [8]), or the oscillation frequency is fixed and the variations of the amplitude and phase are recorded. This mode is commonly named tapping-mode (also known as intermittent contact mode [9]) and is the one that is considered in this study. Tapping-mode (TMAFM) is commonly used because of its ability to probe soft samples, due to the minimization of sample damage during the scanning. Moreover, tapping-mode images can be of two different types: in one type, the image corresponds to the changes of the piezoactuator height necessary to maintain a fixed oscillation amplitude through a feedback loop (the height image); in the other type, the image contains the changes of the oscillator phase delay relative to the excitation signal (phase image). This additional imaging possibility has revealed in many cases a high sensitivity to variations of the local properties. A number of studies have shown the possibility to extract useful information from tapping-mode images of soft samples, especially with samples showing a particular contrast on the local scale, like blends of hard and soft materials [5, 10].

Nevertheless, important questions remain about the physical origin of the image contrast in tapping mode [11-13]. In many cases, the height images are considered to display topographic information, but it must be kept in mind that the local mechanical properties of the samples (*i.e.*, the possibility that the tip slightly penetrates the surface) may also contribute to contrast in the height image. For the phase image, in the dominant repulsive regime, the phase shifts are related to the local mechanical properties. At this point, it is worth mentioning that, in order to maintain the tip in a well-defined oscillating behavior, the perturbation to the oscillator due to the contact with the surface is chosen to be small; in other words, the reduction of the free amplitude (the set-point) is only of a few percent. This method has two advantages: from an experimental point of view, this allows one to identify immediately hard and soft domains, the bright parts of the image corresponding to hard domains. From a theoretical point of view, this allows us to use simple approximations providing analytical solutions able to fit the experimental data [5, 13].

Here, we propose a straightforward method, based on the analysis of approach-retract curves, to provide an unambiguous quantitative measurement of the local mechanical response and/or topographic contribution(s), depending on the studied sample. Moreover, a step further in the understanding of the image contrast is proposed, *via* the analysis of the variation of the phase as a function of the tip-surface distance. The fitting of the experimental data by an appropriate model provides a quantitative evaluation of the contribution of topography, adhesion, indentation and dissipation processes to the contrast. This approach is first applied to thin films of thermoplastic elastomer block copolymers, which are known to phase-separate into well-defined domains on the nanometer scale. The method is extended to other polymer systems, such as polymer blends and polymer nanocomposites.

Experimental Results

Thermoplastic Elastomers as Model Systems

A key point for soft materials is a proper interpretation of the observed contrast. In most cases one has to discriminate between the respective contributions of: (i) the actual topography and (ii) the difference in mechanical properties to the height images. Here, we review a straightforward and easy experimental method to evaluate the contribution of the local mechanical properties to the image contrast. Our approach is based on the reconstruction of height or phase image sections via a rapid analysis of approach-retract curves recorded along those section lines. As recently described [5], a comparison between recorded images and the set of approach-retract curves provides an easy way to discriminate between the topographic and mechanical contributions.

The model system is a thin film of a thermoplastic elastomer, *i.e.*, a block copolymer in which the chemical structure of the sequences is designed in such a way that (i) phase separation of the sequences occurs, giving a well-defined spatial distribution of domains at the nanometer scale and (ii) the domains containing different sequences possess different mechanical properties [5, 6]. Here, the chemical composition has been selected to produce a lamellar morphology. In this case, AFM pictures of a PMMA-b-poly(alkylacrylate)-b-PMMA film present an alternating array of rubbery and glassy lamellae, with a periodicity of 27 nm, as shown in Figure 1.

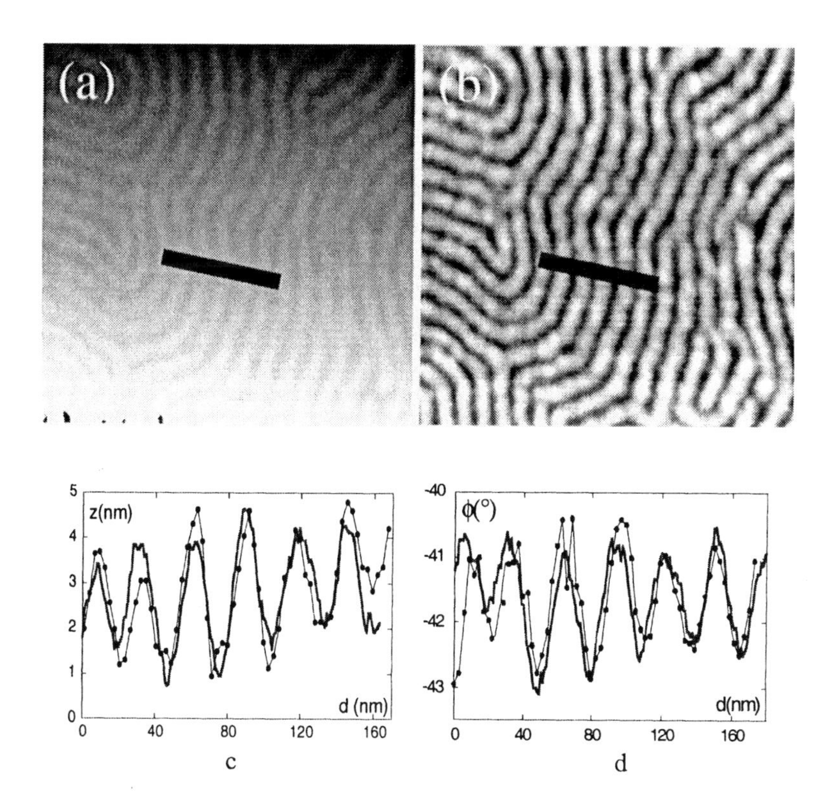

Figure 1. Tapping-mode AFM images (1.0 x 1.0 μm²) of a PMMA-b-poly(alkylacrylate)-b-PMMA film: (a) Height image (b) Phase; Comparison of the image sections with the profiles built from the approach-retract curves data [13]: (c) Height image; (d) Phase image.

The correspondence between the two sets of data (same behavior for the topographic section, not shown here) appears to be very good. This agreement means that for those copolymers, the contrast in the height image is related to different oscillator responses on the glassy and elastomer domains, with no discernible topographic contribution to the contrast. Therefore, the contrast is mostly due to changes in the sample local mechanical properties.

From tapping-mode images of block copolymers, it is not only possible to describe the morphology corresponding to the nanophase separation occurring between the specific domains but also to evaluate accurately the respective contributions of the topography and the mechanical properties. In this case, we can also propose that the phase contrast can be explained on the basis of the viscous forces acting against the tip motion during the indentation of the tip in the sample [13-15].

Applications to other polymer materials.

Biodegradable Polymer Blends. In this section, we extend the concept to polymer blends made of biodegradable and biocompatible components, namely PMMA and poly(ε-caprolactone)(PCL). These blends are in current development for biomedical applications such as drug release systems or prostheses [16]. Figure 2a illustrates the topographic image of a 75:25 weight % PMMA/PCL blend as a thin film. From the section analysis (Figure 2b) and the bearing analysis (not shown here) of the topographic image, it appears that the dark areas, corresponding to the softer component (*i.e.*, the PCL domains), are located approximatively 60 nm below the bright (mechanically harder) zones.

Approach-retract curves recorded on different domains on the polymer blend sample surface are markedly different: the slope of the amplitude/distance curve is 0.75 for the harder, glassy domains while it is smaller (0.60) on the softer domains, meaning that, for the same set-point, there is larger tip indentation in the softer domains. Using the same procedure as described for the block copolymer, we measure a difference of about 15 nm between the lowest points in the image section and the corresponding reconstruction (Figure 2b). Therefore, from these data (160 points), it appears that, in this case, an indentation of about 15 nm in the PCL softer domain has to be considered to fit the experimental data. Note that, at room temperature, the PCL is largely above its glass temperature transition (Tg= -60°C). Therefore, the 60 nm height difference recorded in the height images actually corresponds to a 15 nm indentation into PCL domains that are actually 45 nm below the level of the PMMA domains. This height difference probably originates from the film formation process from the solution of the two polymers in a common solvent. The topographic modulation can be explained by a different rate of solvent evaporation during the film drying process for the two phases [17].

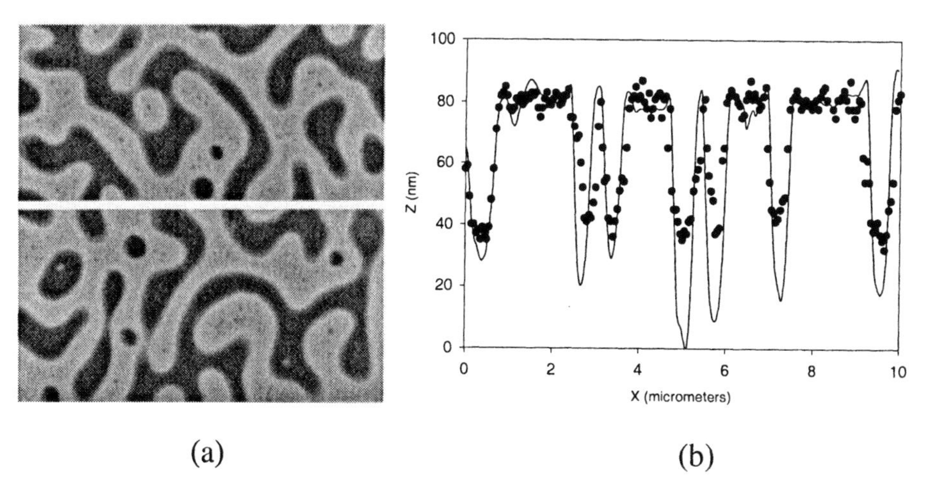

Figure 2. (a) Tapping-mode AFM height image (8.0 x 8.0 μm²) of a PMMA-PCL co-continuous polymer blend (75:25) film. The vertical scale is 80 nm; (b) Image section and its reconstruction from approach-retract curve analysis. The solid line corresponds to the image section and the black dots are the reconstructed data.

Conjugated polymers. We also applied this technique to conjugated organic semiconducting materials. Compared to inorganic materials, conjugated polymers have the advantages of easy control of the semiconducting properties through chemical modification, and ease of processing over large areas, leading to major potential cost savings in device manufacture. LEDs and transistors based on conjugated materials are now being actively developed for commercial applications. Generally, when conjugated oligomers or polymers are deposited from a molecularly dissolved solution on a substrate, like muscovite mica, HOPG or silicon wafers, they tend to form fibrils by self-assembling processes governed by π-stacking of the polymer chains [18-20]. In the present case, the conjugated oligomer is a made of about 20 units of indenofluorene (Figure 3). This polymer is a well-known efficient blue emitter for LED's. Upon annealing at 300°C (above the liquid crystal transition temperature of 270°C), the polymer reorganizes in such a way that the fibrillar morphology is no more present. Figure 4 illustrates the typical morphology observed by AFM after annealing. The analysis of the height image shows a difference of 18 nm between the upper (brighter) layer and the lower (darker) layer. From the phase image (vertical grey scale 1.0 degree) there is no significant contrast between the two layers (only the contours of the cavities are visible).

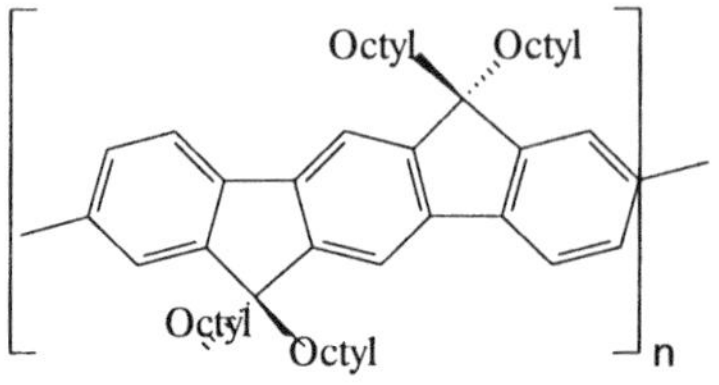

Figure 3. Chemical structure of oligo indenofluorene (n= 20-22).

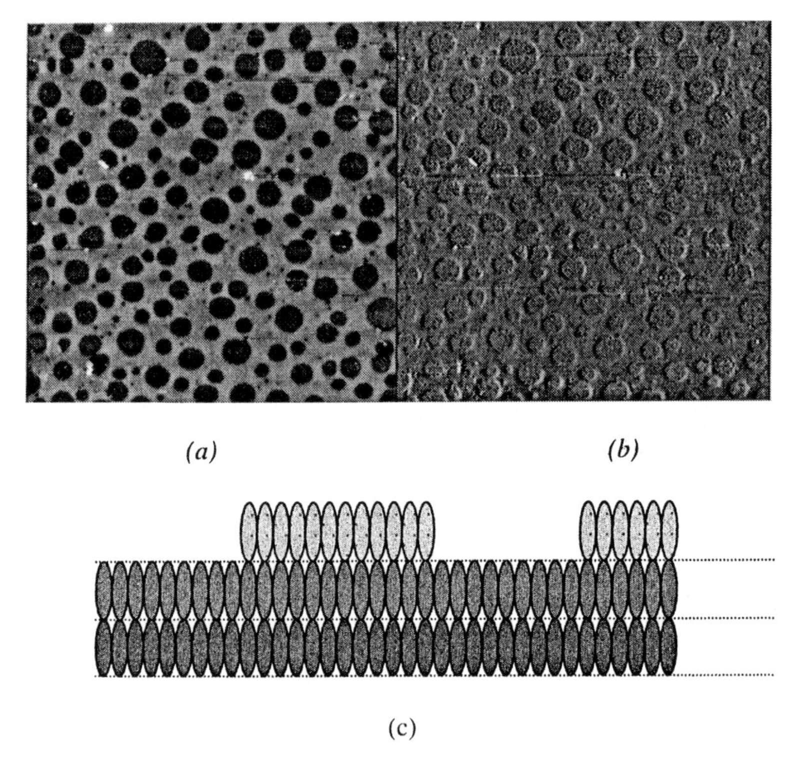

Figure 4. (a) Tapping-mode AFM height image (30.0 x 30.0 μm²) of an oligo indenofluorene thin deposit. The vertical scale is 100 nm; (b) Corresponding phase image. The vertical scale is 1.0 degree; (c) Proposed model for the molecular organization within the film.

The length of the extended indenofluorene oligomer with 20 units is estimated to be around 18.5 nm. This strongly suggests that the layers are made of molecules standing up perpendicularly to the substrate, as depicted in Figure 4c. The fact that the phase contrast between the different layers is very low means that the tip is probing the same material or at least material with the same mechanical properties. For this sample, approach retract curves are strictly identical whatever the location on the sample. This implies that the contrast observed on the height image is purely topographic.

Figure 5 illustrates the morphology adopted by ar conjugated oligomer based on six thiophene units substituted on one end by a short poly(ethylene oxide) segment (Figure 5c). These molecules are deposited on a freshly cleaved HOPG substrate. On the height image, we clearly see a molecular step (indicated by the arrow on Figure 5a). The thickness of brighter islands (indicated by an "A" on Figure 5a) is about 6.0 nm, which roughly corresponds to length of the fully extended molecule. The islands are thus two-dimensional assemblies in which the molecules are perpendicular to the substrate (or slightly tilted). This interpretation is fully confirmed by the phase image, on which the islands show the same phase lag (*i.e.,* the same dark color). Small white spots are also visible (indicated by "C" on Figure 5b). STM analysis of the same sample indicates that this zone is actually the HOPG surface while the light gray zone ("B" on Figure 5b) is made of a monolayer of molecules lying flat. The phase of the oscillating tip is different enough when the tip is over the monolayer, or over an island, or over the HOPG, so that it is possible to understand the nature of the three distinct zones in the AFM image.

Nanocomposites. This work deals with the preparation and the surface characterization of biodegradable nanocomposites made of poly(ε-caprolactone) (PCL) and Montmorillonite-type clay. Nanocomposites with different relative compositions of PCL and Montmorillonite, either natural or organo-modified by various alkylammonium cations, are prepared by melt intercalation and in situ intercalative polymerization [22, 23]. The goal of this study is to characterize the dispersion of the clay layers in the PCL matrix, which is a critical parameter governing the final physical properties of the obtained nanocomposites. Figure 6a shows a TMAFM image of the sample, where flat elongated objects are dispersed in the matrix. On Figure 6b, we can clearly see that the approach curves on zone A and B are drastically different. On zone A, the slope is around 0.99 and is much larger than for zone B (slope = 0.52).

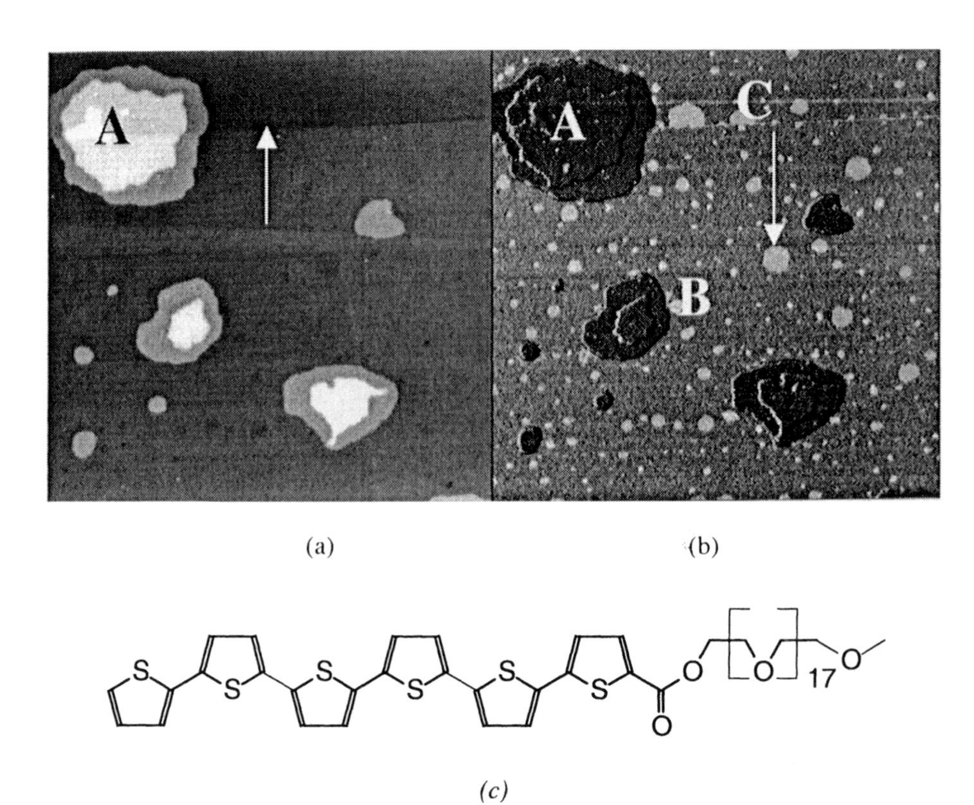

Figure 5. (a) Tapping-mode AFM height image (30.0 x 30.0 μm²) of T6-$(EO)_{18}$ thin deposit on HOPG from THF solution. The vertical scale is 25 nm; (b) Corresponding phase image. The vertical scale is 10 degree; (c) Chemical structure of the oligothiophene molecule.

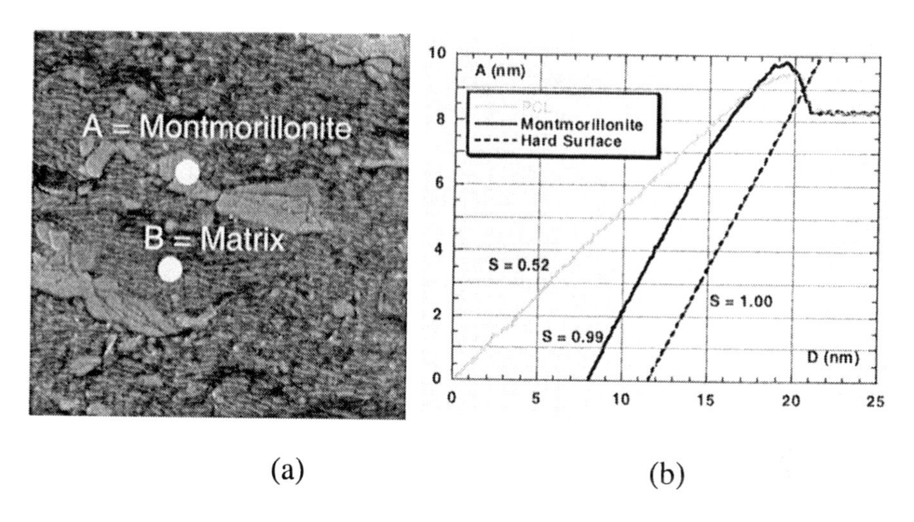

(a) (b)

Figure 6. (a) Tapping-mode AFM phase image (1.0 x 1.0 μm²) of a Montmorillonite-PCL nanocomposite; (b) Experimental approach-retract curves recorded over the PCL matrix and clay inclusions.

A slope value close to unity is the typical signature of a very hard surface (see model curve on Figure 6b), because it corresponds to a situation where no tip indentation takes place. In contrast, a slope value significantly lower than 1 indicates that the tip indents the surface (hence, its oscillation amplitude decreases more slowly than on a hard surface). Thus, we can easily assign the two components from their mechanical responses [18]: zone A is made of Montmorillonite layers while zone B is the PCL matrix. This example illustrates the usefulness of approach-retract curve analysis in the description of the surface morphology.

Conclusions

Here we described a simple and straightforward method (based on the analysis of approach-retract curves) to: (i) give a quantitative estimation of the topographic and the mechanical contribution to height and phase images; (ii) assign the various components in image of heterogeneous polymer materials. This technique has been successively used for pure topographic or pure mechanical contrast as well as for samples with a mixing of the two contributions to the height and phase AFM images.

Acknowledgements

The authors are grateful to J. D. Tong and R. Jérôme (CERM, University of Liège, Belgium) for the synthesis of the triblock copolymers. Conjugated polymers were kindly provided by K. Müllen (MPI-P, Mainz, Germany) and W.J. Feast (IRC Durham, UK). Research in Mons is partially supported by the Belgian Science Policy Program "Pôle d'Attraction Interuniversitaire en Chimie Supramoléculaire et Catalyse Supramoléculaire" (PAI 5/3), the European Commission and the Government of the Région Wallonne (Phasing Out Program). Research in Bordeaux is supported by the Conseil Régional d'Aquitaine. M.S. acknowledges the F.R.I.A. (Belgium) for a doctoral scholarship.

References

1. Leclère, Ph.; Lazzaroni, R.; Brédas, J.L.; Yu, J.M.; Dubois, Ph.; Jérôme, R. *Langmuir* **1996**, *12*, 4317.
2. Stocker, W.; Beckmann, J.; Stadler, R.; Rabe, J.P. *Macromolecules* **1996**, *29*, 7502.
3. Magonov, S.N.; Elings, V.; Wangbo, M.H. *Surf. Sci.* **1997**, *389*, 201.
4. Leclère, Ph.; Moineau, G.; Minet, M.; Dubois, Ph.; Jérôme, R. Brédas, J.L.; Lazzaroni R. *Langmuir* **1999**, *15*, 3915.
5. Kopp-Marsaudon, S.; Leclère, Ph.; Dubourg, F.; Lazzaroni, R.; Aimé, J.P. *Langmuir* **2000**, *16*, 8432.
6. Rasmont, A.; Leclère, Ph.; Doneux, C.; Lambin, G.; Tong, J.D.; Jérôme, R.; Brédas, J.L.; Lazzaroni, R. *Colloids and Surfaces B: Biointerfaces* **2000**,*19*, 381.
7. Konrad, M.; Knoll, A.; Krausch, G.; Magerle, R. *Macromolecules* **2000**, *33*, 5518.
8. Albrecht, T.R.; Grütter, P.; Horne, D.; Rugard D. *J. App. Phys.* **1991**, *69*, 668.
9. Zhong, Q.; Inniss, D.; Kjoller K.; Elings, V.B. *Surf. Sci.* **1993**, *290*, L688.
10. Knoll, A.; Magerle, R.; Krausch, G. *Macromolecules* **2001**, *34*, 4159.
11. Cleveland, J.P.; Anczykowski, B., Schmid, A.E.; Elings V.B. *Appl; Phys. Lett.* **1998**, *72*, 2613.
12. Garcia, R., San Paulo, A. *Phys. Rev. B* **1999**, *60*, 4961.
13. Dubourg, F., Aimé, J.P., Marsaudon, S., Boisgard, R., and Leclère, Ph. *Eur. Phys. J. E*, **2001**, *6*, 49.

14. Dubourg, F., Marsaudon, S., Leclère, Ph., Lazzaroni, R., and Aimé, J.P., *Eur. Phys. J. E*, **2001**, *6*, 387.
15. Leclère, Ph., Dubourg, F., Kopp-Marsaudon, S., Brédas, J.L., Lazzaroni, R., Aimé.J.P., *Appl. Surf. Sc.*, **2002**, *188*, 524.
16. Mayer, J.M. ; Kaplan D.L., *Trends in Polymer* **1994**, *2*, 227.
17. Walheim, S. ; Böltau, M. ; Mlynek J. ; Krausch, G. ; Steiner, U., *Macromolecules,* **1999**, *30*, 4995.
18. Chen, J.T.; Thomas, E.L.; Ober, C.K.; Mao, G.-P. *Science* **1996**, *273*, 343 ; Radzilowski, L.H.; Stupp, S.I. *Macromolecules* **1994**, *27*, 7747; Radzilowski, L.H.; Carragher, B.O.; Stupp, S.I. *Macromolecules* **1997**, *30*, 2110. Jenekhe, S.A.; Chen, X.L. *Science* **1998**, *279*, 1903.
19. Ruokolainen, J.; Mäkinen, R.; Torkkeli, M.; Mäkelä, T.; Serimaa, R.; ten Brinke, G.; Ikkala, O. *Science* **1998**, *280*, 557.
20. Setayesh, S.; Marsitzky, D.; and Müllen K., *Macromolecules* **2000**, *33*, 2016.
21. Henze, O. ; Parker, D. ; Feast, W.J., *J. Mater. Chem.* **2003**, *13*, 1269.
22. Pantoustier, N., Lepoittevin, B., Alexandre, M., Calberg, C., Jérôme, R., Dubois, Ph., *Macromol. Symp.*, **2002**, *183*, 95.
23. Lepoittevin, B., Pantoustier, N., Devalkenaere, M., Alexandre, M., Kubies, D. ; Calberg, C., Jérôme, R., and Dubois, Ph., *Macromolecules*, **2002**, *35*, 8385.
24. Viville, P. ; Lazzaroni, R. ; Pollet, E. ; Alexandre, M. ; Dubois, Ph. ; Borcia, G. ; and Pireaux, J.J. *Langmuir*, **2003**, *19*, 9425.

Chapter 8

Probing Nanorheological Properties of Mesoscopic Polymer Systems

Scott Sills[1], Tomoko Gray[1], Jane Frommer[2], and René M. Overney[1]

[1]Chemical Engineering, University of Washington, Seattle, WA 98195
[2]IBM Almaden Research Center, 650 Harry Road, San Jose, CA 95120

Shear modulation scanning force microscopy (SM-SFM) and lateral force microscopy (LFM) are highlighted as two nanorhelogical tools well suited for exploring the exotic properties of mesoscopic polymer systems. A structural model for rheologically modified interfacial boundary layers is constructed by examining interfacial plasticization, dewetting kinetics, disentanglement barriers, and interfacial glass transition profiles in confined polymer films. Building on this foundation, the utility of SM-SFM and LFM in technological development is demonstrated with the novel process of ultrahigh density data storage in confined polymer films.

Mesoscale Interfacial Confinement

The laws of nature, or more appropriately, their perception in the form of physiochemical material properties, become challenged when a three-dimensional bulk material is confined to only two dimensions. This is apparent in many interfacial applications, such as thin film technologies. Interfacial technologies are trapped between the atomistic and the three-dimensional bulk regimes – in the mesoscale. Here, the phenomenological classifications applied macrosystems breakdown. On the other hand, quantum or molecular theories are insufficient in describing the fractal-like behavior of mesoscopic systems. Interfacial constraints generate frustrated, or poorly mixed, systems, in which equilibration processes are severely impeded. These systems are best described with non-equilibrium statistical mechanical models for constrained small ensemble systems.

Miniaturization trends of electronic, optical, mechanical, and biomedical devices are bringing the mesoscale to the forefront of the engineering design arena. The challenge for continued evolution of mesoscale technologies is to work within the interfacial constraints, or yet, to utilize these constraints as engineering design opportunities. Progress relies heavily on the development of appropriate analytical techniques. Over the last decade, scanning force microscopy (SFM) has adapted to meet some of these needs.

In this chapter, shear modulation SFM (SM-SFM) and lateral force microscopy (LFM) are introduced as nano-rheological tools. With these tools, a variety of rheological behaviors are explored in confined polymeric systems, and a structural model for interfacial boundary layers is constructed. Once the foundation of interfacial nanorheology is established, the utility of SM-SFM and LFM in technological development is exemplified with the novel SFM based process of ultrahigh density data storage (Tb/in^2) in confined polymer films (*1*).

Shear Modulation – Scanning Force Microscopy (SM-SFM)

Shear modulated SFM (SM-SFM) is a non-scanning nano-rheological method that has proved particularly successful in determining crosslinking densities and structural phase transitions, e.g. glass transitions, of ultra-thin films (*2, 3*). The method involves a nanometer sharp SFM cantilever tip that is maintained at constant load in contact with a sample, as illustrated in Figure 1.

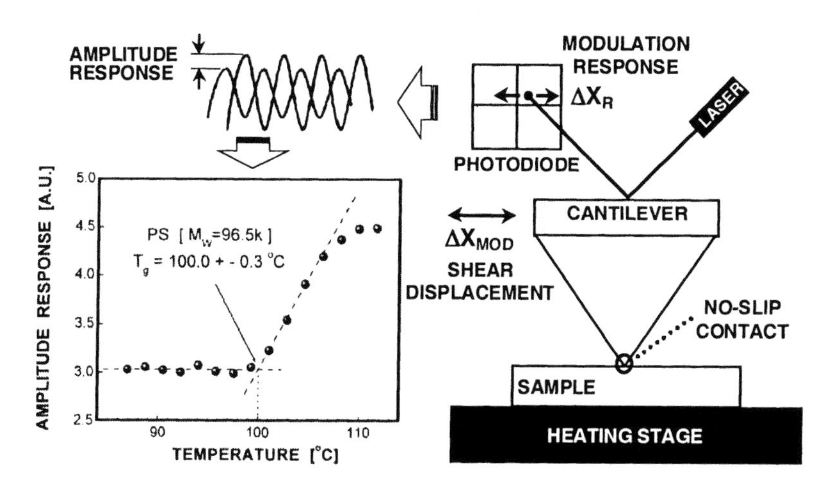

Figure 1: Shear modulated scanning force microscopy (SM-SFM). The T_g on a thick polystyrene film is indicated by the "kink" in the amplitude response.

The tip is laterally modulated with a nanometer amplitude (ΔX_{MOD}) that avoids any tip-sample slipping. The modulation response (ΔX_R) is analyzed relative to ΔX_{MOD} using lock-in techniques. The response amplitude is a measure of the contact stiffness (i.e. sample modulus) (*3*), and is useful for tracking thermo-rheological transitions. For example, the glass transition temperature (T_g) of polystyrene is indicated at the *kink* of the response curve in Figure 1.

Below the T_g, the probing depth of the SM-SFM is on the order of 1nm, which allows *substrate-independent* measurements down to film thicknesses of a few nanometers. Any surface effects less than 1nm in depth (*4*) cannot be addressed under these conditions. The slow creeping process above T_g is documented elsewhere (*5*). While the accuracy of SM-SFM T_g measurements compares well with other techniques (*6*), SM-SFM also offers the versatility for probing rheological properties in confined sample geometries.

Lateral Force Microscopy (LFM)

Lateral Force Spectroscopy (LFM) simulates a single asperity provided by an ultra-sharp tip on a soft cantilever. The small contact area, on the order of the molecular dimensions, is insufficient to coherently reorganize macromolecules and allows discussing LFM results in terms of thermodynamic equilibrium. The working details of LFM are outlined in (*7*), and calibration methods for determining the absolute friction force, F, are discussed in (*8*). It is important that friction analyses on compliant materials are *adhesion corrected,* shown in Figure 2, to account for load dependent changes in the contact area, i.e. F must be described in terms of the normal force (F_N) which is the sum of the applied load (F_{APP}) and the adhesion force (F_{ADHS}).

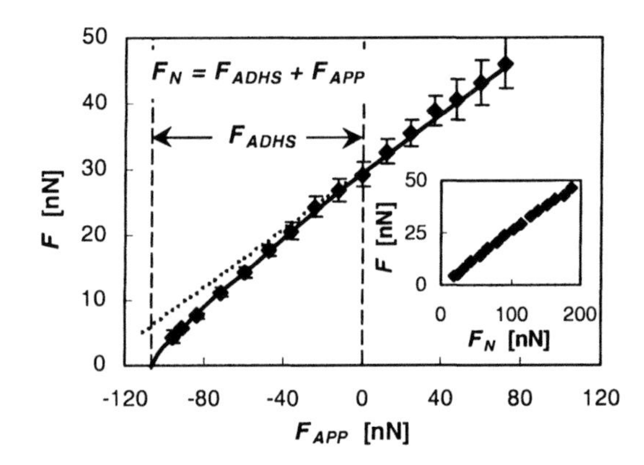

Figure 2: Adhesion correction of friction measurements on a polymer film.

LFM offers sufficient sensitivity for friction measurements on the molecular scale, and has been used to detect molecular stick-slip behavior (*9*). Until recently, the question remained of whether or not idealized theoretical tribology models, e.g. creep models (*10-12*), would apply to polymeric systems. LFM studies on glassy, atactic polystyrene (PS) (*13*) offered the first known experimental confirmation of creeping friction dynamics in amorphous polymers. Friction-rate-temperature measurements revealed logarithmic friction-velocity relationships that correspond with LFM experiments on ionic crystals (*14*) and in lubricated sliding (*15*). These data are well described by creeping friction models (*10-12*), and suggest that the dissipative process is consistent with barrier-hopping fluctuations on a periodic surface potential.

The potential barrier (E) is determined by using the method of reduced variables (*16*) to superpose the $F(v)$ isotherms into a master curve. The superposed data along with the a_T shift factor are shown in Figure 3. A representative barrier height of 7.0 kcal/mol is deduced from the Arrhenius behavior of a_T. This value corresponds to the activation energy for phenyl ring rotation around the C-C bond with the backbone chain, 7.0 kcal/mol (*17*).

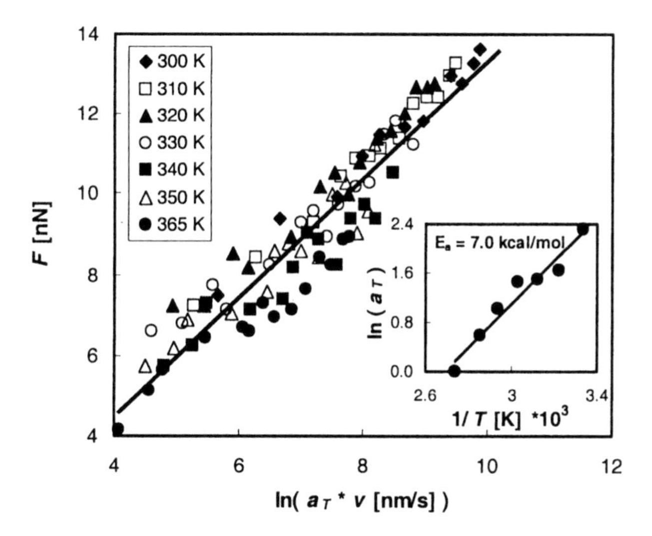

Figure 3: Superposed master curve for $F(v)$ isotherms on PS, and (inset) the thermal a_T shift factor indicating an activation engergy(E) of 7.0 kcal/mol.

Considering that the creep models (a) are based on a Gaussian fluctuation distribution (*10-12*) and (b) fit the friction data reasonably well, one may conclude that there is little or no correlation between the individual phenyl rotations that are *relaxed* during the sliding motion. It is because of this week correlation that similar frictional behaviors are observed on both highly structured (crystalline) surfaces and the unstructured (amorphous) PS surface.

LFM studies have provided valuable insight to the classical field of tribology. However, the use of LFM in probing polymer relaxations as dissipation mechanisms incorporates a rheological component into the tribological scenario. Polymer relaxation processes are often characterized by monomeric, or internal, friction. In this light, the tribo-rheological LFM analysis provides a bridge between internal rheological friction and external tribological friction; two phenomena generally considered irrespective from one another.

Interfacial Nanorheology: Substrate Effects in Polymer Films

The discussion thus far has focused on SFM rheological techniques and bulk-property measurements. In polymer thin films, when film thicknesses approach the nanometer scale, structural, material, and transport properties become increasingly dominated by interfacial and dimensional constraints. Rheological boundary layers are often formed at interfaces, within which, anisotropic constraints lead to bulk-deviating behaviors. This section is devoted to exploring rheological boundary layers at polymer interfaces. A variety of rheological manifestations are illustrated with several LFM and SM-SFM studies, and a visualization of the molecular configuration within interfacial boundaries is gradually developed with each successive example.

Interfacial Plasticization

The conceptually intuitive process of heterogeneous diffusion serves as worthy starting point for a discussion about rheologically modified interfacial boundary layers. This is illustrated for a multiphase binary thin film system, in which, low molecular weight components (LMCs) leach from an underlying film into the surface film, forming an interdiffusion zone at the interface. LFM studies were conducted on poly(methyl methacrylate) (PMMA) films supported on either crosslinked epoxy or Si substrates.

For PMMA films on epoxy substrates, the friction coefficient (μ) in Figure 4 decreases with increasing film thickness, and for thicker films, approaches the friction value of PMMA on Si. The friction coefficient may be considered constant on all films, if one assumes that (a) the shear modulus is constant, and (b) the adhesion force is only a function of the contact area, i.e. a constant physical and chemical bonding strength between the LFM tip and sample. Holding to these assumptions and considering that the LFM probe (Si) is much stiffer than PMMA, it follows that changes in μ reflect changes in the PMMA modulus.

The friction coefficient for the 35nm films is significantly higher with epoxy substrates than for Si. However, the friction coefficient on thicker epoxy supported films reaches the low value found on the 35nm Si supported film, appearing substrate independent. This friction gradient suggests the leaching of LMCs from the epoxy into the PMMA that is illustrated in Figure 4, essentially *softens,* or *plasticizes,* the film and reduces the PMMA modulus.

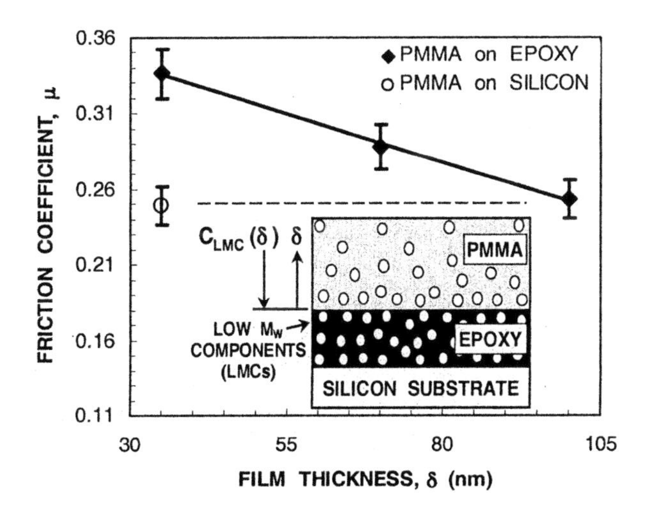

Figure 4: LFM friction coefficient measurements of PMMA supported on Si and crosslinked epoxy substrates disclose the interfacial plasticization process. (inset) model for low molecular weight component leaching into PMMA.

The extent of the softening, and modulus depression, is proportional to the concentration of LMCs (C_{LMC}) at the surface, which in turn, is a function of film thickness. For a thickness of 100nm, the friction coefficient matches that of the 35nm Si supported PMMA, indicating no detectable plasticization, or LMCs, at the surface. This process of heterogeneous interdiffusion across interfaces highlights the importance of substrate chemistry for thin film applications. In this case, a 100nm thick boundary layer is rheologically modified due to the plasticization effects of interdiffused low molecular weight components.

Dewetting Kinetics

The prior discussion of interfacial plasticization served as one example where chemical transport processes are responsible for the formation of a rheologically modified interfacial boundary. The rheological impact associated with physical and dimensional constraints is perhaps less intuitive. Never the less, various groups have reported bulk-deviating structural and dynamic properties for polymers at interfaces (*18-22*). For example, reduced molecular mobility in ultra-thin PS films was reported based on forward recoil spectroscopy measurements (*18*).

The role of the substrate in generating rheologically modified boundary layers becomes apparent in dewetting studies with binary PS on polyethylene-co-propylene (PEP) films, which are supported on Si substrates (high interaction surface) (*19*). The dewetting kinetics in Figure 5 were determined from a time-series of SFM topography images, and reveal a critical PEP film thickness, δ_{CRIT}, below which, the dewetting velocity (V_d) decreases with decreasing film thickness, and above which, V_d remains constant. Independent LFM measurements on Si supported PEP films also indicate a critical film thickness, δ_{CRIT}, below which, the friction decreases with decreasing film thickness, and above which remains constant. In both studies, the critical PEP film thickness in Figure 5 corresponds to approximately 100nm.

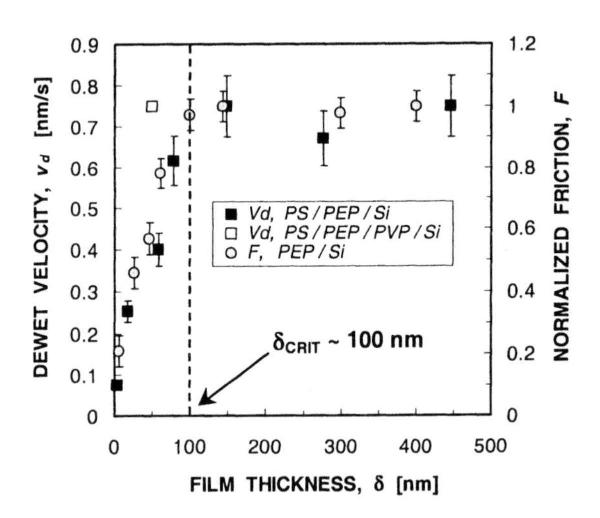

Figure 5: Dewetting velocity (V_d) and friction (F) measurements on PS/PEP systems reveal a 100nm interfacial boundary layer. Data from reference (19).

The dewetting kinetics and friction forces both suggest the presence of a rheologically modified PEP boundary layer adjacent to the Si. For $\delta_{PEP} < \delta_{CRIT}$, the decreasing friction represents an increase in the PEP modulus. This translates to an increasing *glasslike* behavior, or loss of mobility, as the Si interface is approached through the PEP phase. It is this loss of PEP mobility that is responsible for decreasing the dewetting velocity.

To identify the source of this rheological gradient, the PEP-Si interactions were effectively masked by first spin casting a low interaction foundation layer of poly(vinyl pyridine) (PVP) on the Si. The dewetting velocity of the PS/PEP/PVP film is reported as the open box in Figure 5 and remained constant, even at PEP film thicknesses below δ_{CRIT}. This anomalous finding unveils the high interfacial interactions between PEP and Si as responsible for the apparent PEP vitrification inside the interfacial boundary.

Disentanglement Barriers

The current picture of the rheological boundary attributes formation of the boundary layer to interfacial constraints on the molecular mobility. In the past, interfacial effects were considered to be confined to the pinning regime, typically on the order of a few nanometers. However, LFM disentanglement studies on PEP films (*23*) and NMR tracer diffusion measurements in PS (*21*) have revealed that the interfacial boundary may extend up to 10 radii of gyration (R_G) beyond the interface. Simple surface pinning alone has been ruled out since, at this distance, the probability of a polymer molecule making direct surface contact is nearly zero.

LFM friction measurements on Si supported PEP films (R_G =24nm) offer insight to the source of these *far-field* molecular constraints. A transition in the friction coefficient at a critical load (P) is seen in Figure 6. The higher friction coefficient below P portrays a dissipative behavior consistent with viscous plowing through an entangled PEP melt.

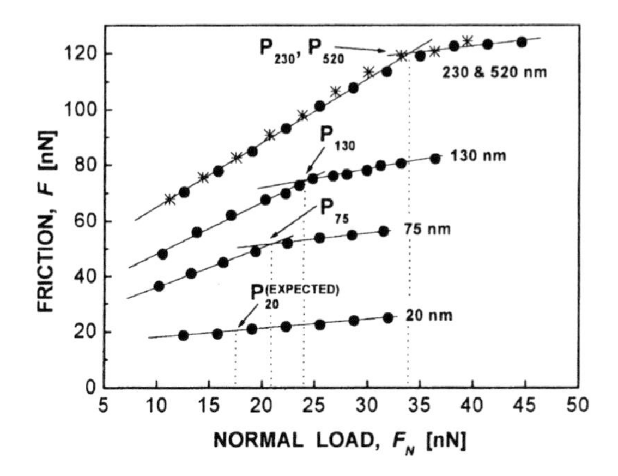

Figure 6. LFM measurements on PEP films reveal a critical load (P) marking a transition from viscious shearing to chain sliding. (Data are from reference 23. Copyright 1999 American Chemical Society.)

At loads exceeding *P*, the reduced friction coefficient represents a chain slipping phenomenon similar to a shear banding behavior. Thus, the critical load may be conceptualized as an effective activation barrier for disentanglement (*23*). The boundary layer thickness and information about to the conformational structure within the boundary are elucidated from the film thickness dependence of *P*:

(i) The absence of the disentanglement transition (*P*) in the 20nm films and the ubiquitous low friction, chain slipping suggest that the PEP molecules are highly disentangled within a *sublayer* immediately adjacent to the substrate.

(ii) In the 75-230nm films, the disentanglement transition (*P*) increases linearly with film thickness until the bulk-material *P* is reached. The sub-bulk *P* values indicate an *intermediate regime* of partial disentanglement, the extent of which diminishes with increasing film thickness until the bulk entanglement density is recovered. This far-field disentanglement (~ 10 R_G from the substrate) is attributed to the strain imposed during spin casting, and the preservation of the disentangled structure in the melt reflects an anisotropic diffusion process where partially disentangled chain ends diffuse into the porous structure (*21*) of sublayer (*23*).

(iii) Finally, for films thicker than 230nm, the polymer behaves like the bulk elastomer and loses any memory of the underlying Si.

The picture of rheological boundary layers now reflects a two phase system comprised of a *sublayer* and an *intermediate regime.* The mobility constraints are ascribed to the strain imposed during spin casting, paired with interfacial interactions in the sublayer and anisotropic diffusion in the intermediate regime.

Interfacial Glass Transition Profiles

One parameter of particular interest when discussing molecular mobility of polymers is the glass transition temperature (T_g). For thin homopolymer films, it has been recognized that several factors are intricately responsible for the departure of T_g from the bulk value (*3-5, 24-28*), e.g. the proximity of a free surface, substrate interactions, and process-induced anisotropy. Here, we address the effects of spin casting on the interfacial T_g profiles of amorphous polymer films. In addition, the use of chemical crosslinking as a controlled mobility constraint in thin films is investigated, and the rheological impact is discussed.

SM-SFM T_g measurements on 12kDa PS films are presented in Figure 7. For film thicknesses (δ) > 200nm, the T_g values correspond to the bulk T_g of 95°C (*6*). A two phase boundary layer is encountered within ~200nm of the substrate: (a) T_g values are depressed relative to the bulk in a *sublayer* with a thickness on the order of $R_{G,}$ one order of magnitude beyond the persistence length (*29*); and (b) T_g values exceed the bulk T_g in the *intermediate regime.*

This non-monotonic $T_g(\delta)$ relationship is interpreted considering two competing processes that affect the relaxation dynamics: (a) shear induced structuring and (b) interdiffusion (*21, 23, 30*). Shear structuring creates an interfacial region where the spin casting shear stresses induce polymer stretching and or disentanglement (structural deformation). The second process involves the interdiffusion between the entropically cooled interfacial region and the unperturbed bulk phase.

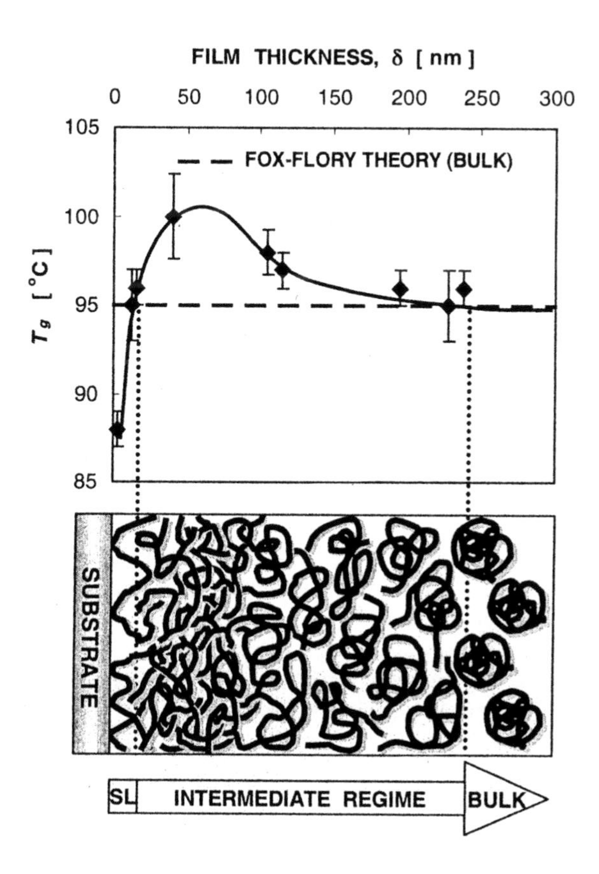

Figure 7: (top) Film thickness (δ) dependence of T_g for 12kDa PS films compared to the bulk T_g from Fox-Flory theory. (bottom) Rheological boundary model for the observed $T_g(\delta)$ relation (SL=sublayer). (Data used with permission from reference 31. Copyright 2004 American Institute of Physics.)

With a strong precedence for shear structuring in PS solutions (*32-34*) and considering the shear stress profiles during spin casting (*35*) it is reasonable to propose that the effects of spin casting extend from the substrate to the boundary with the bulk phase. In this scenario, the extent of structural deformation is related to the shear stress profile during casting. Alternatively, the shear structuring may extend only through the sublayer, and interdiffusion alone may be responsible for the conformational restructuring in the intermediate regime. For this case, the molecular mobility is limited by the propagation of *holes*, or packets of free volume, which facilitate conformational rearrangements (*36*).

Spin casting films of increased molecular weight (M_w) had the effect of shifting the $T_g(\delta)$ profiles further from the substrate, by ~10nm/kDa (*31*). The bulk T_g is recovered at ~250nm for all films in the 12-21kDa M_W range. The influence of M_w on the internal structure of the boundary layer appears more pronounced on the sublayer thickness than on the far-field boundary of the intermediate regime. This suggests that the overall boundary thickness depends more strongly on the spin casting shear stresses than on molecular dimensions.

When the molecular weight is increased by crosslinking pre-cast PS-BCB films, the $T_g(\delta)$ profiles exhibit a similar qualitative behavior before and after crosslinking (*31*), indicating a preservation of the rheological anisotropy after crosslinking at 250°C, ~150°C above T_g. The crosslinking yields an overall T_g increase of 7±3°C; however, in contrast to the M_W dependence discussed above, no spatial shift is found in the $T_g(\delta)$ profiles. Since crosslinking occurs after spin casting, the shear stresses that create the shifted $T_g(\delta)$ profiles are not present. Hence, the $T_g(\delta)$ profiles are impacted differently for each condition of increased M_W because of the sequence of treatments.

The two phase model for rheological boundary layers has evolved to include interfacial interactions that lead to the formation of a less dense *sublayer* adjacent to the interface. The thickness of the sublayer is characterized, in part, by the molecular dimensions and the interaction potential at the interface. The coupled effects of shear-induced structuring during spin casting and anisotropic relaxation and transport constraints during annealing are responsible for the creation of an *intermediate regime* between the sublayer and bulk phase. The overall rheological boundary may extend up to two orders of magnitude beyond the polymer's persistence length, and the molecular restructuring within the boundary is thermally stable well above T_g. Finally, the impact of mobility constraints (e.g. crosslinking) on the structure within boundary layers depends on the sequence of the film preparation process, i.e. constrains incorporated before and after casting exhibit different rheological outcomes.

Scanning Probes in Thermomechanical Data Storage

With a precedence established for bulk-deviating behaviors in confined polymer systems, the development and optimization efforts of thin film technologies will be challenged to work within these interfacial constraints, or yet, to utilize the constraints as engineering design opportunities. Very specific material engineering may be achieved through an understanding of polymer dynamics at the polymer-substrate interface. Modified relaxational properties and enhanced conformational stability may be achieved through control of the molecular weight, crosslinking density, and film thickness. In this regard, the characterization and control of interfacial boundary layers becomes increasingly important.

As the LFM and SM-SFM techniques led, in part, to the discovery and visualization of rheological boundary layers, these nanorheological tools will continue to support the evolution of mesoscale technologies. Ultrahigh density (Tb/in^2) thermomechanical data storage (*1*) is one such example that is particularly relevant to the discussion of scanning probes. Briefly described, data are stored thermomechanically with SFM cantilever type probes, and relies on the writing, reading, and erasing of nanometer sized datum-bit *indentations* in ~50nm polymer films, pictured in Figure 8

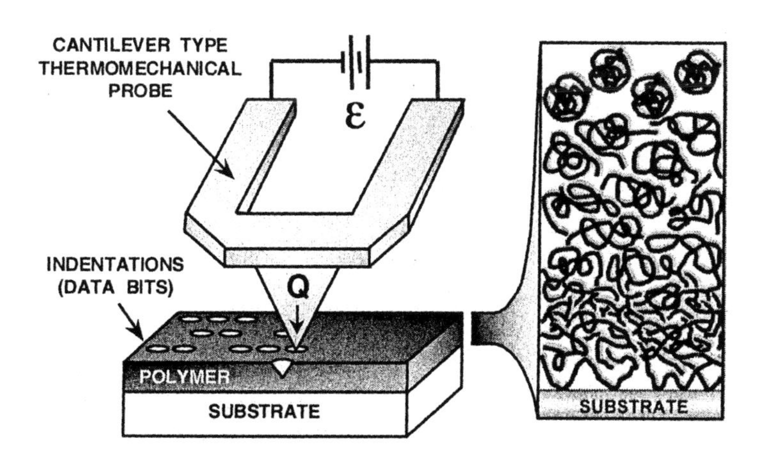

Figure 8: Thermomechanical data storage in thin polymer films(~50nm) must operate with the molecular constraints present in the interfacial boundary layer.

The ideal polymer medium should be easily deformable for bit writing; however, the written bits must be stable against wear, thermal degradation, and dewetting. In addition, the polymer must allow reversible bit indentations, i.e. controlled backflow, for erasing and re-writing. Anticipating the effects of interfacial confinement on the thermo-mechanical properties within the rheological boundary pictured in Figure 8 presents one challenge to optimizing the polymer storage media. Ultimately, achieving the desired performance goals will require engineered polymeric media *of tailored molecular structure* in order to achieve the narrow range of physiochemical properties necessary for: low power consumption, high durability, and long shelf life. Development of this technology continues to benefit from various LFM and SM-SFM investigations, similar to many of the studies highlighted here.

Acknowledgements

The following individuals are gratefully acknowledged for their contribution to this work: Cynthia Buenviaje at Intel; Wilson Chau, Victor Lee, Craig Hawker, Brooke van Horn, Bob Miller, and Chuck Wade at IBM Almaden; and Bernd Gotsmann, Urs Duerig, and Peter Vettiger at IBM Rueschlikon. Funding for this work was provided in part by IBM; the ACS Petroleum Research Fund; the University of Washington Center for Nanotechnology; the National Science Foundation, NSF MRSEC (DMR96324235), CTS-9908912; and the Royalty Research Fund of the University of Washington.

References

1. Vettiger, P.; Cross, G.; Despont, M.; Drechsler, U.; Duerig, U.; Heberle, W.; Lantz, M. I.; Rothuizen, H. E.; Stutz, R.; Binnig, G. K. *IEEE Transactions on Nanotechnology* **2002**, *1*, 39-55.
2. Luginbuhl, R.; Overney, R. M.; Ratner, B. D. In *Interfacital Properties on the Submicron Scale;* Frommer, J.; Overney, R. M.; ACS Symposium Series 781; Oxford University Press: Oxford, England, 2000; pp 179-96.
3. Overney, R. M.; Buenviaje, C.; Luginbuhl, R.; Dinelli, F. *J. Therm. Anal. Cal.* **2000**, *59*, 205-25.
4. de Gennes, P.G. *C.R. Acad. Sci.* **2000**, *IV*, 1-8.
5. Ge, S.; Pu, Y.; Zhang, W.; Rafailovich, M.; Sokolov, J.; Buenviaje, C.; Buckmaster, R.; Overney, R. M. *Phys. Rev. Lett.* **2000**, *85*, 2340-3.
6. Buenviaje, C.; Dinelli, F.; Overney, R. M. In *Interfacital Properties on the Submicron Scale;* Frommer, J.; Overney, R. M.; ACS Symposium Series 781; Oxford University Press: Oxford, England, 2000; pp 85.
7. Overney, R.; Meyer, E. *MRS Bulletin* **1993**, *XVIII*, 26-34.
8. Buenviaje, C. K.; Ge, S. R.; Rafailovich, M. H.; Overney, R. M. *Mat. Res. Soc. Symp. Proc.* **1998**, *522*, 187-92.
9. Overney, R. M.; Takano, H.; Fujihira, M.; Paulus, W.; Ringsdorf, H. *Phys. Rev. Lett.* **1994**, *72*, 3546-49.
10. Dudko, O. K.; Filippov, A. E.; Klafter, J.; Urbakh, M. *Chem. Phys. Lett.* **2002**, *352*, 499-504.
11. Sang, Y.; Dube, M.; Grant, M. *Phys. Rev. Lett.* **2001**, *87*, 174301/1-4.
12. Heslot, F.; Baumberger, T.; Perrin, B.; Caroli, B.; Caroli, C. *Phys. Rev. E* **1994**, *49*, 4973-88.
13. Sills, S. E.; Overney, R. M. *Phys. Rev. Lett.* **2003**, *91*, 095501/1-4.
14. Gnecco, E.; Bennewitz, R.; Gyalog, T.; Loppacher, C.; Bammerlin, M.; Meyer, E.; Güntherodt, H.-J. *Phys. Rev. Lett.* **2000**, *84*, 1172-5.

15. He, M.; Szuchmacher Blum, A.; Overney, G.; Overney, R. M. *Phys. Rev. Lett.* **2002**, *88*, 154302/1-4.
16. Ferry, J. D. *Viscoelastic Properties of Polymers,* 3rd ed.; John Wiley: New York, 1980; pp 264-318.
17. Reich, S.; Eisenberg, A. *J. Poly. Sci. A-2* **1972**, *10*, 1397-400.
18. Frank, B.; Gast, A. P.; Russel, T. P.; Brown, H. R.; Hawker, C. *Macromolecules* **1996**, *29*, 6531-4.
19. Overney, R. M.; Guo, L.; Totsuka, H.; Rafailovich, M.; Sokolov, J.; Schwarz, S. A., In *Dynamics in Small Confining Systems IV*; Drake, J. M.; Klafter, J.; Kopelman, R; Mat. Res. Soc.: Pittsburgh, PA, 1997; Vol. 464, pp 133.
20. Rabeony, M.; Pfeiffer, D. G.; Behal, S. K.; Disko, M.; Dozier, W. D.; Thiyagarajan, P.; Lin, M. Y. *J. Che. Soc. Faraday Trans.* **1995**, *91*, 2855-61.
21. Zheng, X.; Rafailovich, M. H.; Sokolov, J.; Strzhemechny, Y.; Schwarz, S. A.; Sauer, B. B.; Rubinstein, M. *Phys. Rev. Lett.* **1997**, *79*, 241-4.
22. Overney, R. M.; Leta, D. P.; Fetters, L. J.; Liu, Y.; Rafailovich, M. H.; Sokolov, J. *J. Vac. Sci. Technol.* **1996**, *B 14*, 1276-9.
23. Buenviaje, C.; Ge, S.; Rafailovich, M.; Sokolov, J.; Drake, J. M.; Overney, R. M. *Langmuir* **1999**, *15*, 6446-50.
24. Chen, H. P.; Katsis, D.; Mastrangelo, J. C.; Chem, S. H.; Jacobs, S. D.; Hood, P. J. *Advanced Materials* **2000**, *12*, 1283-6.
25. Dalnoki-Veress, K.; Forrest, J. A.; de Gennes, P. G.; Dutcher, J. R. *J. Phys. IV* **2000**, *10*, 221-6.
26. Forrest, J. A.; Mattsson, J. *Phys. Rev. E* **2000**, *61*, R53-R61.
27. Tseng, K. C.; Turro, N. J.; Durning, C. J. *Phys. Rev. E* **2000**, *61*, 1800-11.
28. Kim, J. H.; Jang, J.; Zin, W. C. *Langmuir* **2001**, *17*, 2703-10.
29. Brogly, M.; Bistac, S.; J., S. *Macromol. Theor. Simul.* **1998**, *7*, 65-8.
30. Orts, W. J.; Van Zanten, J. H.; Wu, W. L.; Satija, S. K. *Phys. Rev. Lett.* **1993**, *71*, 867.
31. Sills, S. E.; Overney, R. M.; Chau, W.; Lee, V. Y.; Miller, R. D.; Frommer, J. *J. Chem. Phys.* **2004**, *in press.*
32. Saito, S.; Hashimoto, T. *J. Chem. Phys.* **2001**, *114*, 10531-43.
33. Endoh, M. K.; Saito, S.; Hashimoto, T. *Macromolecules* **2002**, *35*, 7692-9.
34. Morfin, I.; Linder, P.; Boue, F. *Macromolecules* **1999**, *32*, 7208-23.
35. Wu, X.; Squires, K. D. *J. Fluid Mech.* **2000**, *418*, 231-64.
36. Wang, C. L.; Wang, S. J. *Phys. Rev. B* **1995**, *51*, 8810-4.

Chapter 9

Combined Atomic Force Microscopy and Sum Frequency Generation Vibrational Spectroscopy Studies of Polyolefins and Hydrogels at Interfaces

Aric Opdahl and Gabor A. Somorjai

Department of Chemistry, University of California, and Materials Sciences Division, Lawrence Berkeley National Laboratory, Berkeley, CA 94720

AFM and sum frequency generation (SFG) vibrational spectroscopy experiments provide complementary information that can be used to relate the morphology and mechanical properties of a polymer interface to the molecular structure of the interface. The application of the two techniques to the study of polymer interface structure is presented, focusing on surface segregation and wetting behavior of polyolefin blends and on the interface structure and mechanical behavior of hydrogels exposed to various hydration conditions.

Introduction

This chapter emphasizes the incorporation of sum frequency generation (SFG) vibrational spectroscopy to AFM studies of polymer behavior at interfaces. In recent years, AFM has been proven as a versatile technique that can be used to map the surface morphology/topography and to probe the mechanical properties of polymer interfaces in a variety of environments. In spite of this versatility, one of the major limitations of AFM is the lack of direct measurement of the molecular structure of the interface. Although chemical information can sometimes be inferred from the interface mechanical properties (elasticity, adhesion), or by experiments where the AFM tip is coated with a specific chemical functionality, AFM experiments in general lack chemical specificity.

SFG spectroscopy experiments fill this void by providing information regarding the molecular structure of the interface. SFG surface vibrational spectroscopy is an optical technique that is highly sensitive to the chemical composition, orientation, and ordering of molecular groups at an interface. Like AFM, SFG spectroscopy experiments can be designed to probe polymer/air and polymer/liquid interfaces, making this a powerful combination of experimental techniques for *in situ* studies of interface phenomenon.

Two sets of experiments that highlight the application of both AFM and SFG spectroscopy to the study of polymer interface behavior are presented in this chapter. The first focuses on surface segregation and wetting behavior of polyolefin copolymers and blends. In that example, chemical information obtained by SFG spectroscopy is connected to lateral morphology measurements made of the interface by AFM. The second example focuses on the interface behavior of hydrogels. That example highlights the versatility of both techniques for studying polymer/liquid interfaces and in connecting mechanical measurements of the interface made by AFM to chemical measurements made by SFG spectroscopy.

Experimental

SFG surface vibrational spectroscopy

Excellent descriptions of SFG surface vibrational spectroscopy have been published by Shen[1,2,3,4] and by Hirose[5]. The application of SFG surface vibrational spectroscopy to the study of polymer interfaces has recently been reviewed.[6] In our experiments, SFG vibrational spectra of polymer/air and polymer/liquid interfaces were obtained by overlapping two laser beams at the interface and measuring the light generated from the interface at the sum frequency in the reflected direction. One of the input fields is a tunable infrared beam that is resonant with one or more vibrational modes of the species at an interface. The picosecond laser and OPG/OPA system we have used to generate the visible beam (ω_1, 532 nm) and the tunable infrared beam (ω_2, 2000 cm^{-1} to 4000 cm^{-1}) has been described elsewhere.[7]

The interface specificity of SFG spectroscopy is a result of selection rules obtained under the electric-dipole approximation. The sum frequency signal, $I(\omega_s)$, is proportional to the square of the nonlinear susceptibility of the material being measured, $\ddot{\chi}^{(2)}$, a 27 component tensor (Eq. 1). Under the electric-dipole approximation, the 27 components of $\ddot{\chi}^{(2)}$ are equal to zero for centrosymmetric materials. Materials that are centrosymmetric, or that are randomly oriented in the bulk, are not expected to generate large sum frequency signals. However, if a material assumes a preferred orientation at an interface – then symmetry is broken in the interface plane, and some of the components of $\ddot{\chi}^{(2)}$ may be non-zero. Measurement of $\ddot{\chi}^{(2)}$ is specifically sensitive to this type of ordering at an interface.

$$I(\omega_s) \propto \left|\chi^{(2)}\right|^2 \qquad (1)$$

The vibrationally resonant contribution to the nonlinear susceptibility, $n_s\langle\ddot{\alpha}_R^{(2)}\rangle_f$ is enhanced when the infrared beam (ω_2) is tuned near a vibrational mode belonging to one of the molecular groups at the interface (ω_q). The molecular hyperpolarizability, $\alpha^{(2)}$, can be related to the product of the dynamic dipole and polarizabilities of a vibrational mode. Thus, the mode must be both IR and Raman active in order to be measured. The measured strength, $\ddot{A}_q$, for a particular vibrational mode, q, is directly proportional to the number density of contributing molecular groups at the surface, and the orientation averaged nonlinear polarizability of those groups.

$$\ddot{\chi}^{(2)} = \ddot{\chi}_{NR}^{(2)} + n_s\langle\ddot{\alpha}_R^{(2)}\rangle_f = \ddot{\chi}_{NR}^{(2)} + \sum_q \frac{\ddot{A}_q}{\omega_2 - \omega_q + i\Gamma_q} \qquad (2)$$

Surface vibrational spectra presented in this chapter were obtained in the CH (2700 cm^{-1} - 3100 cm^{-1}) stretching region using the $s_{sum}s_{vis}p_{IR}$ polarization combination. This polarization combination is most sensitive to molecular groups that have vibrational modes with a component of the IR transition moment oriented normal to the surface plane and a component of the polarizability tensor in the surface plane. SFG spectra were fit using Eq. 1 and 2 in order to extract peak positions and amplitudes. The measured $\ddot{A}_q$ can be used to estimate the number density and orientation of the ordered molecular groups giving rise to the vibration in the interface region - if the components of $\ddot{a}_q$ for a vibrational mode are known. Methods for estimating the magnitude of the components of $\ddot{a}_q$ for C-H vibrational modes are presented in references 2 and 5.

AFM

Topographic and friction images, as well as measurements of mechanical properties, were collected using a commercial AFM capable of imaging large areas (100 μm x100 μm) and a homebuilt, walking-style AFM scanning head. The homebuilt AFM is completely enclosed within a glass bell jar. Relative humidity (RH) in the bell jar can be varied by balancing the evaporation of water from a reservoir with a steady flow of nitrogen through the chamber. Decreasing the flow rate of nitrogen increases the experimental humidity, which is measured by a hydrometer placed within the chamber.

Polymers

Low and high molecular weight atactic polypropylene (*a*PP, M_w ~50,000 and M_w ~200,000), aspecific polyethylene-co-propylene rubber (*a*EPR, M_w ~50,000; 42 mol% ethylene randomly incorporated), and isotactic polypropylene (*i*PP, M_w ~200,000) were synthesized by Basell Polyolefins Inc. Further description of the physical properties of these polymers is available in references 7 and 8. Further description of the poly(hydroxyethyl)methacrylate (pHEMA) hydrogels, provided by Ocular Sciences, Inc. is given in reference 9.

Surface segregation in polyolefin copolymers and blends

This section summarizes SFG and AFM experiments designed to probe surface segregation and wetting behavior of polyolefin blends. The blend components include atactic polypropylene (*a*PP), isotactic polypropylene (*i*PP), and aspecific poly(ethylene-co-propylene rubber) (*a*EPR)[8,10]. In this polymer system, AFM experiments distinguish *a*PP and *a*EPR by their differences in viscoelastic properties, and are used to characterize lateral morphology (phase separation) at the interface. The iPP component is distinguished by its crystallinity. SFG experiments can be performed to determine the surface composition and configuration of the components at the interface. We have previously used this strategy to characterize the surface composition of bulk-miscible biomedical polyurethane blends, and it was shown that the component with lowest surface energy migrated to the polymer/air interface.[11]

Blends of *a*PP, *i*PP, and *a*EPR serve as mimics for the important commercial blend of isospecific PP/EPR. A practical problem of PP/EPR blends is that they typically have poor adhesive properties – this is a characteristic of many crystalline polyolefins. The adhesive properties of the PP/EPR surface, however, have been observed to vary depending on the bulk composition and the processing conditions; therefore, it is important to understand how the surface composition is affected by both of these variables. Although polymer/air interfaces of bulk miscible (single phase) polymer blends are usually enriched in the component, which has the lowest surface energy, most polymer blends are immiscible (multi-phase). The surface morphology of immiscible blends is complex, and complete wetting (surface segregation) of the lower surface tension component is not always observed.[12]

Atactic polypropylene (*a*PP)/air and aspecific poly(ethylene-*co*-propylene) rubber (*a*EPR)/air interfaces

SFG spectra (ssp polarization) obtained from two of the blend components, *a*PP and *a*EPR, are shown in Figure 1. These spectra are specifically sensitive to the number density and orientation of ordered molecular groups in the interface region, and are uniquely sensitive to the polymer configuration at the polymer/air interface. In each of these spectra, the strongest resonant feature is at 2883cm^{-1} and is assigned as the CH_3(s) stretch from the methyl side branches.[13,7] The feature at 2968cm^{-1} is assigned to the CH_3(a) stretch from the side branch. The features at 2850cm^{-1} and 2920cm^{-1} are assigned as the CH_2(s) and CH_2(a) stretches, respectively, from the polymer backbone.

Like molecules in liquids,[3] polymers have generally been observed by SFG to preferentially orient the lowest surface energy structural unit at the polymer/air interface.[6] The strong CH_3(s) peak suggests that the lower surface energy methyl side branches have a tendency to order at polymer/air interfaces. In contrast, for hydrocarbon polymers without short side branches (e.g. polyethylene[13] and polyethylene glycol (PEG)[14]) it has been observed that backbone CH_2 units preferentially order at the polymer/air interface. For both CH_2 and CH_3 molecular group, the features associated with the CH_2(s) and CH_3(s) vibrations will be strongest in the ssp spectra if the methyl groups are oriented upright (with C_{3v} axis normal to the surface plane). A large ratio of the CH_3(s)/CH_3(a) features peaks observed in the *a*PP and *a*EPR spectra is also an indication of an upright orientation.

An important observation in the SFG spectrum of *a*EPR is the small reduction in absolute intensity of the peak associated with the CH_3(s) vibration, as compared to its intensity in the spectrum of *a*PP. The *a*EPR copolymer is comprised of ~60% propylene monomers randomly incorporated in the polymer

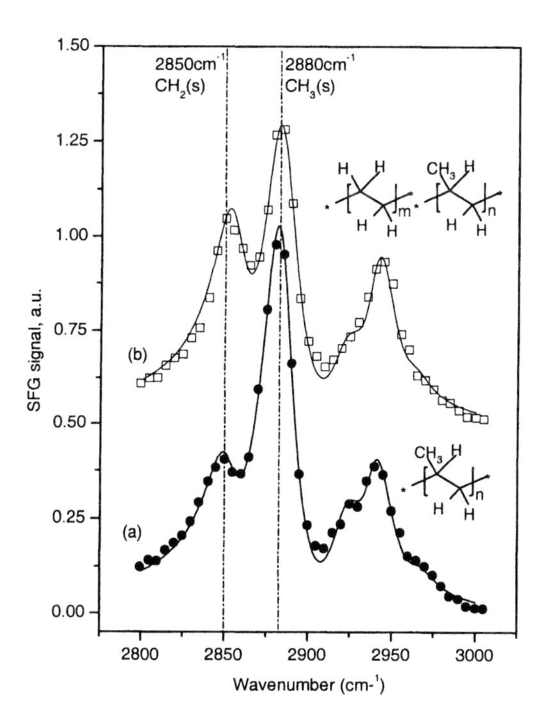

Figure 1. SFG spectra of (a) aPP/air and (b) aEPR/air interfaces ($s_{sum}s_{vis}p_{IR}$ polarization. (Reproduced from reference 10. Copyright 2002 American Chemical Society.)

chain, thus a given chain length has ~60% as many methyl side branches as *a*PP. Since the SFG signal intensities scale as n^2, the peak associated with the CH_3(s) vibration from *a*EPR is predicted to be only ~36% as intense as that from *a*PP (assuming the surfaces of both materials reflect the bulk concentration). That the measured *a*EPR CH_3(s) peak is much larger suggests at least three possibilities: (1) a higher fraction of methyl side branches are ordered at the *a*EPR surface compared to the *a*PP surface (2) the methyl side branches are oriented with the C_{3v} symmetry axis more upright at the *a*EPR/air interface than they are at the *a*PP/air interface, or (3) the methyl side branches are more tightly ordered for *a*EPR than they are for *a*PP. Analysis of the SFG spectra indicated that the most likely scenario is (1) and that ethylene-rich *a*EPR copolymers orient excess methyl branches at the polymer/air interface compared to *a*PP.[7]

This result from *a*PP and *a*EPR can be put into perspective with SFG results obtained from other relatively simple polymers - polystyrene[15,16,17,18], poly(hydroxyethyl)methacrylate,[9] and polymethylmethacrylate[19] where it has been observed that bulky low surface energy side branches tend to order at the polymer/air interface. In comparing these results to results from liquid interfaces, one must take into account a major difference between polymers and molecules in liquids - that the configuration of a polymer chain is restricted by the connectivity of chain segments along the backbone.

Polymers do not have the conformational freedom that molecules in liquids have, and consequently the equilibrium surface structure likely reflects a compromise between ordering a particular structural unit at the interface and conformational restrictions imposed by the polymer chain. The *a*PP and *a*EPR SFG spectra suggest that the number of side branches oriented at the polymer/air interface may be limited by this types of restriction imposed by the chain architecture.

*a*PP/*a*EPR blends

The surfaces of two amorphous blends of *a*PP/*a*EPR are compared in this section - a bulk miscible and a bulk immiscible blend.[10] SFG spectra obtained from a low molecular weight *a*PP (Mw 50,000)/*a*EPR blend (50:50 wt. percent) is shown in Figure 2a. The spectra were obtained immediately after casting from n-hexane and after an extended annealing period. Differential scanning calorimetry (DSC) results indicate that this blend is miscible in the bulk. Qualitatively, the SFG spectrum obtained from the blend surface after the extended annealing period is nearly identical to the spectrum obtained from the surface of *a*PP – indicating preferential segregation of *a*PP to the surface. Figure 2b presents results obtained from an *a*PP/*a*EPR blend where the molecular weight of the *a*PP component was increased from 50,000 to 200,000 – where

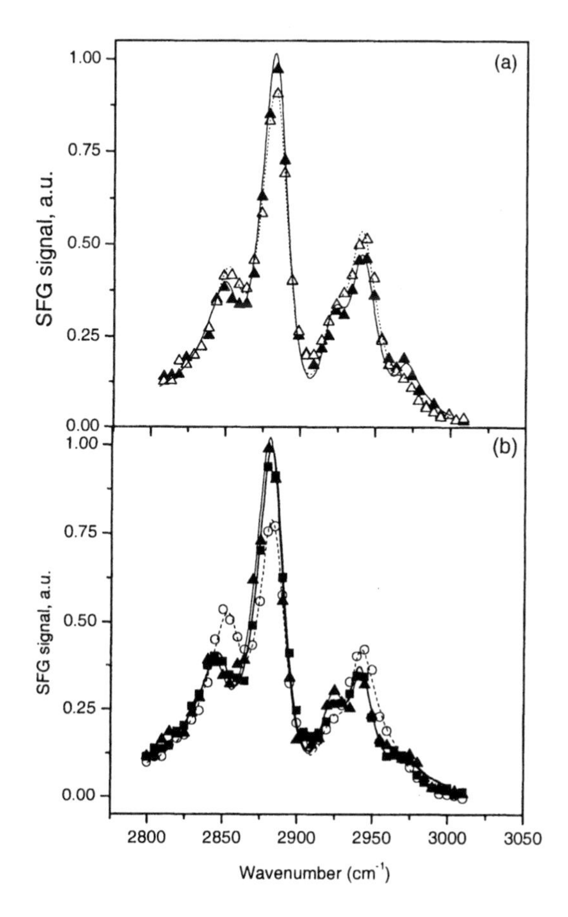

Figure 2.(a) SFG spectra of low molecular weight aPP/aEPR blend "as cast" (open triangle) and after annealing period (filled triangle). (b) SFG spectra of high molecular weight aPP (filled triangle), aEPR (open circle), and 50:50 bulk immiscible aPP/aEPR blend (filled square). (Reproduced from reference 10. Copyright 2002 American Chemical Society.)

DSC results showed that the blend becomes immiscible in the bulk. The behavior for this blend is similar to the miscible blend. In both miscible and immiscible *a*PP/*a*EPR blends, SFG spectra indicate that *a*PP segregates to the blend/air interface at the monolayer level.

For comparison, we have also characterized these surfaces by XPS.[10]Like SFG spectroscopy, XPS results obtained from the carbon valence band region also distinguish between *a*PP and *a*EPR based on differences in CH_3 content. In contrast though, the XPS experiements have a surface sensitivity that is determined by the mean free path of photoelectrons generated in the polymer film (in this particular case ~5-7 nm - much deeper than the monolayer sensitivity of SFG experiments). XPS experiments were unable to detect significant surface enrichment of *a*PP in the bulk miscible *a*PP/*a*EPR blend. In connection with the SFG data, this suggests that the *a*PP surface enrichment layer in the miscible blend is very thin, and that the highest levels of *a*PP in the surface enrichment layer are restricted to the top 2-3nm of the film. In contrast, XPS spectra obtained from the bulk immiscible blend did show significant enrichment of *a*PP in the surface region, suggesting that the *a*PP enrichment layer for the immiscible blend is thicker than 5-7nm. Thus it is concluded that for *a*PP/*a*EPR blends, that *a*PP wets the surface of the blend and that decreasing the miscibility increases the thickness of the surface enrichment layer.

AFM data support this conclusion. AFM images were used to monitor the wetting process in the *a*PP/*a*EPR blend. AFM images obtained from the miscible low molecular weight *a*PP/*a*EPR blends are homogeneous at all stages of annealing. AFM topography and lateral force images obtained from the surface of the bulk immiscible blend at various stages of the annealing process, are shown in Figure 3. Immediately after casting, the surface of the immiscible blend is rough. The lateral force image shows the presence of 'low friction' (dark) and 'high friction' (light) phases. The *a*EPR has a much lower elastic modulus than the high molecular weight *a*PP component, thus the 'high friction' regions are assigned as *a*EPR. As the blend is annealed, the surface becomes smoother and the 'high friction' *a*EPR regions become covered by an *a*PP wetting layer.

iPP/aEPR blends

When isotactic polypropylene (*i*PP) is substituted for atactic polypropylene in the PP/*a*EPR blend, blend surfaces can be prepared that are enriched in *i*PP, enriched in *a*EPR, or that contain a phase separated mix of the two components. This is due to the crystalline nature of the isotactic component, which can trap morphologies formed in the melt into a nonequilibrium conformation. Distinguishing between *i*PP and *a*EPR by AFM is straightforward as the *i*PP component is crystalline and the *a*EPR component is soft and amorphous. Figure

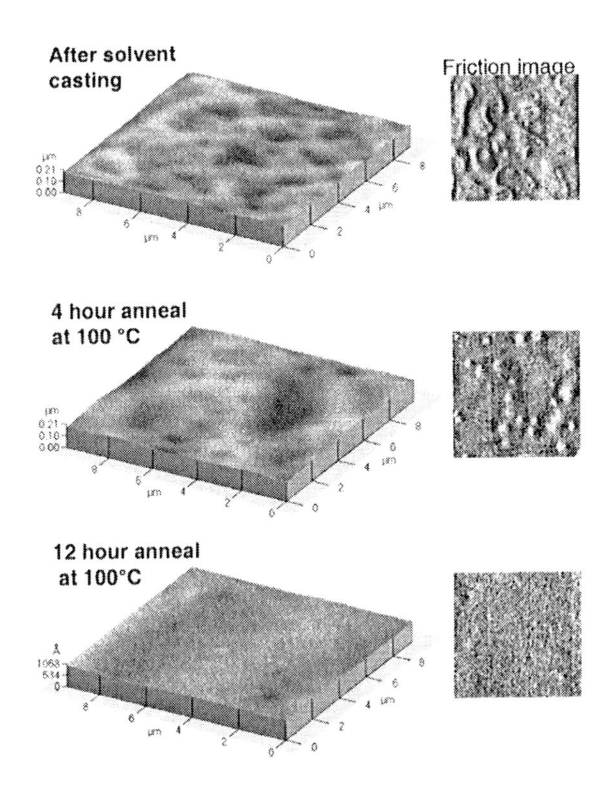

Figure 3.AFM images of bulk immiscible aPP/aEPR blend at various annealing stages.

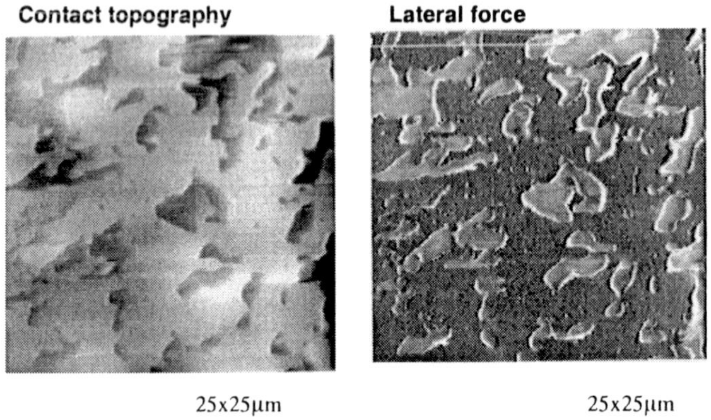

Contact topography

Lateral force

25x25μm

25x25μm

Figure 4. AFM topography and lateral force images of melt pressed iPP/aEPR blend. (Reproduced with permission from reference 8. Copyright 2004.)

4 shows an AFM topography and corresponding friction image of the *i*PP/*a*EPR blend melt pressed between glass for 30 seconds and quenched to room temperature. The depressed regions (dark) in the topography image correspond to high friction (bright) regions in the friction image and have been assigned as *a*EPR. In this image, the depressions are an artifact caused by the tip pressing against the soft *a*EPR phase. Scanning with higher load, the tip presses deeper into the surface of the blend.

When the blend is melt-pressed for longer times, the *i*PP and *a*EPR phases grow in size, but are both present at the surface after the pressing substrate is removed. When the blend is melted in open air, the surface appears crystalline by AFM. This suggests that *i*PP preferentially segregates to the air/polymer interface – consistent with the results obtained from the *a*PP/*a*EPR blends. For the iPP/aEPR samples, the surfaces of the air melt surfaces are too rough to obtain high-quality SFG spectra from, however, XPS data and the AFM images suggest that the segregation of iPP is thick in this case.

After melt-pressing, the *a*EPR component can be enriched at the interface by exposing the blend to n-hexane solvent vapor. The *a*EPR component is soluble to n-hexane and the *i*PP component is not. The solvent swells the *a*EPR and draws it to the interface. This is seen in AFM images obtained from *i*PP/*a*EPR surface exposed to n-hexane vapor, which appear amorphous and are mechanically softer after treatment with n-hexane vapor. The *a*EPR has better adhesive properties than the *i*PP component, and in practical applications drawing the EPR component to the surface with solvent may improve the overall adhesive properties of the PP/EPR blend.

Surface molecular structure and surface mechanical properties of hydrogels, and adsorption at the polymer/liquid interface

Hydrogels have been used as soft contact lenses for vision correction for over 30 years. In spite of the many advances that have been made to improve the comfort and biocompatibility of contact lenses, the interfacial properties of contact lens hydrogels, including surface hydration, are not well understood.[20] It is generally believed that high water content and high surface hydrophilicity are desirable properties, in order to increase the wettability of tear films.[21] Surface water content is particularly important for poly(hydroxyethyl)methacrylate (pHEMA) based contact lenses, which tend to dehydrate and become glassy and rigid when they are on the eye.[22] Additionally, when a contact lens hydrogel is placed on the eye, protein material adsorbs to the lens from the tear fluid, which eventually leads to discomfort for the wearer.

AFM and SFG experiments can be designed to measure properties relevant to these types of problems at both the hydrogel/air and hydrogel/liquid interfaces.[9,23,24,25] A schematic of the specially designed AFM sample holder used to study both types of interfaces is shown in Figure 5. AFM measurements were made on fully hydrated hydrogels at the hydrogel/water interface, fully dehydrated (no saline in the reservoir) hydrogels at the hydrogel/air interface, and on hydrogels exposed to controlled humidity air at the hydrogel/humidified air interface. SFG and AFM experiments were performed in a similar manner, but with SFG using flat hydrogel samples instead of curved contact lens hydrogels. The surfaces of two hydrogels were compared: (a) a neutral pHEMA hydrogel (38% wt water when fully hydrated) and (b) an ionic hydrogel comprised of a copolymer of pHEMA and methacrylic acid (55% wt water when fully hydrated).

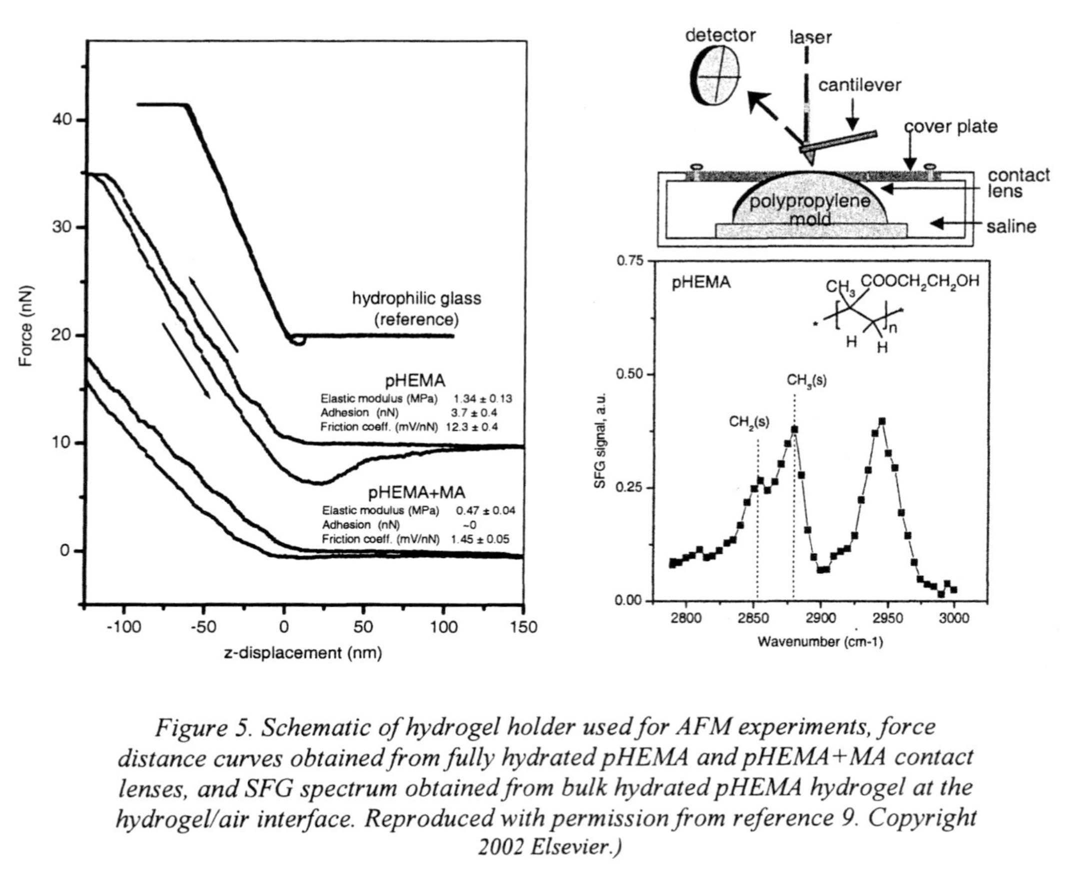

Figure 5. Schematic of hydrogel holder used for AFM experiments, force distance curves obtained from fully hydrated pHEMA and pHEMA+MA contact lenses, and SFG spectrum obtained from bulk hydrated pHEMA hydrogel at the hydrogel/air interface. Reproduced with permission from reference 9. Copyright 2002 Elsevier.)

Surface molecular structure and mechanical behavior of hydrogels

An SFG spectrum obtained from the hydogel/air interface of a bulk hydrated pHEMA contact lens is shown in Figure 5. This spectrum contains features associated with the hydrophobic methyl branch and the hydrophilic hydroxyethyl branch. In contrast, SFG spectra obtained from dehydrated hydrogels at the hydrogel/air interface show significantly less contribution from the hydrophilic side branch while SFG spectra obtained from the 'swollen' hydrogel/water interface are generally featureless.[9] This type of behavior was also observed for polystyrene that was exposed to toluene (solvent) vapor.[26] Toluene is a solvent for polystyrene and will readily penetrate a polystyrene film at ambient temperature. SFG spectra obtained from the toluene liquid/vapor interface showing that in addition to coating the surface of polystyrene, toluene penetrates and disrupts the ordering of the underlying polystyrene film. The results for pHEMA suggest that, like the hydrophobic polystyrene/toluene interface, the hydrophilic hydrogel is not highly ordered at the hydrogel/water interface.

AFM indentation (force vs. distance) curves collected from the surfaces of fully hydrated pHEMA and pHEMA+MA hydrogels are shown in Figure 5. For reference, a force vs. distance curve collected against a hydrophilic silica surface is also shown in Figure 5. These indentation curves measure the bending of the AFM cantilever as the AFM tip presses against the hydrogel surface and is then retracted from the hydrogel. The slopes of the approach and retract curves contain information related to the stiffness, elastic modulus, and viscoelastic relaxation of the hydrogel surface.

A large (1-μm radius of curvature) polystyrene AFM tip was used to produce the indentation curves shown in Figure 5. This tip applies lower pressure to the hydrogel than a conventional 20-nm radius of curvature tip. Qualitatively, the slope of the pHEMA+MA curve is smaller than the slope of the pHEMA curve. The lower slope obtained from the pHEMA+MA surface suggests a lower stiffness and elastic modulus for the pHEMA+MA hydrogel surface – consistent with the lower bulk modulus of this higher water content hydrogel. Applying the Hertz contact model, estimates of the elastic modulus for the fully hydrated lenses are consistent with the measured bulk values. Additionally, the adhesion (pull-off force) measured between the polystyrene AFM tip and the hydrogel surface is much higher for the neutral pHEMA hydrogel than it is for the ionic pHEMA+MA lens. In aqueous solution, this type of behavior is associated with hydrophobic interactions occurring at the interface between polystyrene and pHEMA.[27]

Adsorption at the polymer/liquid interface

To further understand hydrophobic interactions, we have used SFG spectroscopy to investigate the interface structure of other amphiphilic neutral polymers, polypropylene glycol (PPG), polyethylene glycol (PEG), and a triblock PEG-PPG-PEG copolymer, adsorbed on hydrophobic polystyrene surfaces at the solid/liquid interface.[28] It is commonly believed that the driving force for the adsorption of macromolecules to hydrophobic surfaces in aqueous solution is the entropy change associated with the removal of ordered water from the hydration shell of the macromolecules and the substrates.[29,30,31,32] In addition to aiding in the understanding of the high adhesion measured between polystyrene and pHEMA, these types of experiments are important for understanding the basic mechanisms of adsorption and denaturation of proteins at interfaces.[33]

SFG spectra in the CH stretch region obtained from amphiphilic neutral PPG, PEG, and a PEG-PPG-PEG copolymer adsorbed on hydrophobic perdeuterated polystyrene(PS) and hydrophilic silica substrates from aqueous solution are shown in Figure 6a and 6b, respectively. The use of perdeuterated polystyrene removes contributions from the substrate in the C-H stretching region of the SFG spectra.

When adsorbed on hydrophobic polystyrene surfaces, the SFG spectrum for PPG contains features around 2840, 2870, 2940, and 2970 cm^{-1}, assigned to the $CH_2(s)$, $CH_3(s)$, CH_3 Fermi resonance ($CH_3(F)$), and $CH_3(a)$ modes, respectively.[34] PEG is more hydrophilic and has a higher solubility in water, compared to PPG. The spectrum of PEG adsorbed to polystyrene contains vibrational features around 2865 cm^{-1} and 2935 cm^{-1}, corresponding to the $CH_2(s)$ and CH_2 Fermi resonance ($CH_2(F)$) modes, respectively.[35] These two SFG spectra suggest that adsorbed PPG and PEG molecules order at the hydrophobic polystyrene/water interface. The SFG spectrum for a triblock copolymer (PEG-PPG-PEG) adsorbed on hydrophobic polystyrene surfaces contains features similar to the spectrum obtained from PPG. This suggests that the more hydrophobic PPG center block orders, producing an SFG signal, whereas the more hydrophilic PEG end blocks are disordered when the triblock copolymer adsorbs on the hydrophobic polystyrene.

To test the role that the hydrophobic solid surface plays in amphiphilic neutral polymer ordering, a similar set of experiments was conducted using bare silica substrates in place of polystyrene. Clean silica surfaces are generally regarded as hydrophilic because of the presence of surface silanol groups (Si-OH) and can be prepared as described elsewhere.[36] SFG spectra in Figure 6b show no spectral features for any of the polymers adsorbed at the silica/water interface, despite large amounts of polymer adsorption measured upon removal of the substrate from the solution. This indicates that when amphiphilic neutral

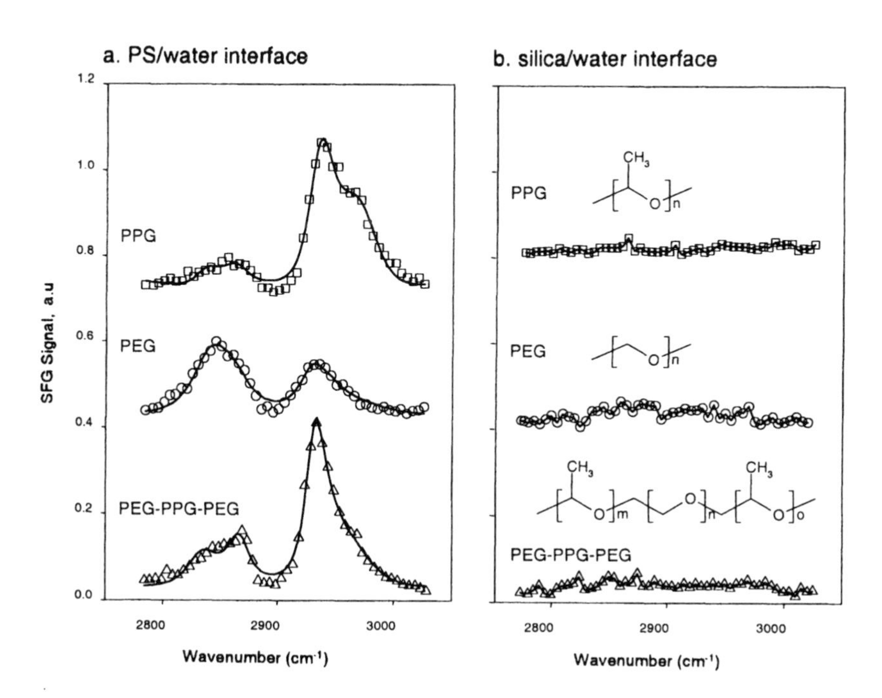

Figure 6. SFG spectra ($s_{sum}s_{vis}p_{ir}$) obtained at the solid/liquid interface of PEG, PPG, and PEG-PPG-PEG adsorbed to (a) a hydrophobic perdeuterated polystyrene surface and (b) a hydrophilic silica surface. (Reproduced from reference 28. Copyright 2003 American Chemical Society.)

polymers like PPG, PEG, and PEG-PPG-PEG adsorb on hydrophilic surfaces, they do not preferentially orient their hydrophobic or hydrophilic moieties.[37]

To demonstrate the role of water in the alignment of adsorbed polymers at the interface, a set of experiments was performed using deuterated methanol in place of water as a solvent. SFG spectra obtained from the hydrophobic polystyrene surfaces and hydrophilic silica surfaces in contact with PPG/methanol solutions contained no measurable features, indicating that polymers adsorbing on the substrates do not display ordering.[38] Similar results were obtained for PEG and the triblock PEG-PPG-PEG copolymer.

These results indicate that hydrophobic surfaces along with water as a solvent are required for the ordering of adsorbed amphiphilic neutral polymers. This may indicate that part of the high adhesion between pHEMA and polystyrene correlates to ordering of the polymer (and loss of water) at the pHEMA/polystyrene interface. Charged polymers do not show ordering at polystyrene. This can explain the lower adhesion between the hydrogel and the tip as well as the lower friction coefficient (obtained from friction vs. load lines[9]) measured for the ionic pHEMA+MA hydrogel compared to the pHEMA hydrogel surface.

Humidity dependence of hydrogel surface mechanical properties

In an ocular enviroment, the bulk of a hydrogel contact lens is hydrated and the surface of the lens is exposed to air (when the eye is open). This situation is mimicked by the experimental setup shown in Figure 5, where the hydrogel surface is exposed to a controlled humidity enviroment. In this experiment, AFM force-distance indentation curves were obtained over a range of rates for different set humidity values. The measurement of rate-dependent surface mechanics provides information concerning the viscoelastic properties and relative hydration of the hydrogel surface as a function of humidity.

Figure 7 shows two AFM indentation curves obtained at different probing rates from the surface of a bulk hydrated pHEMA hydrogel at 75% relative humidity, using a conventional sharp (20-nm radius of curvature) silicon nitride AFM tip.[23] For the curve obtained at a slower probing rate, the cantilever bends less (the slope of the approach curve is smaller), indicating that for slower loading rates, the AFM tip presses deeper into the hydrogel as the polymer chains relax. Relaxation of the pHEMA hydrogel surface can be qualitatively assessed by measuring the slope of the approach curve as a function of probing rate. Figure 7c shows the dependence of the approach curve slope on the probing rate for humidity values between 45% and 80%.

At low humidity, there is little rate dependence in the indentation curves. The slopes of the indenting (approach) curves measured from bulk-hydrated

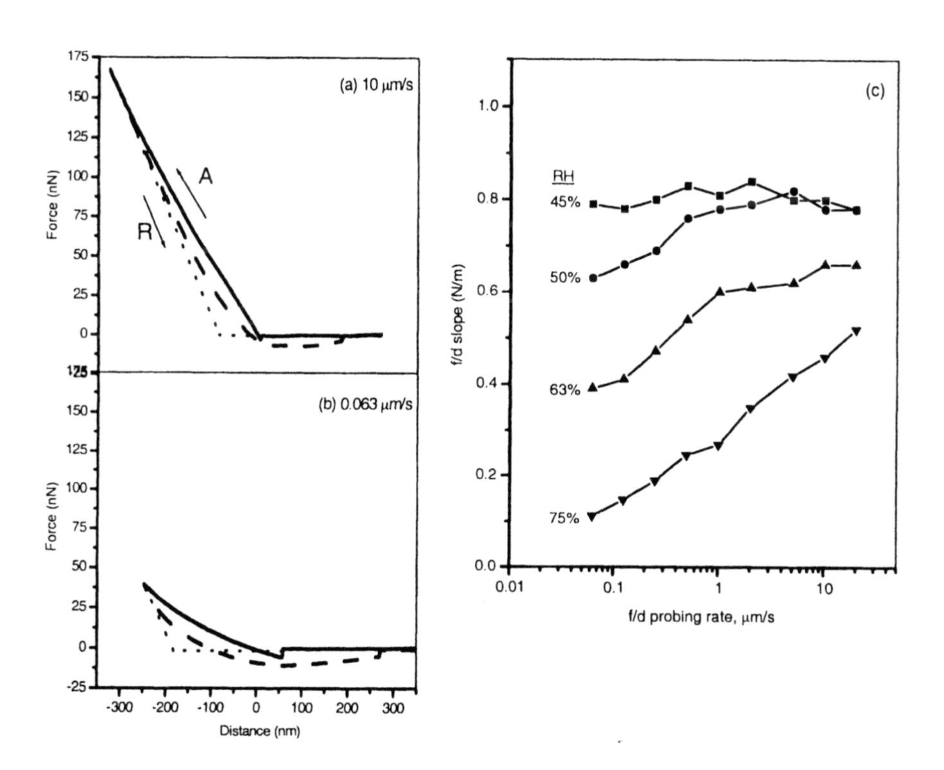

Figure 7. AFM force vs. distance indentation curves obtained from a pHEMA hydrogel exposed to 75% relative humidity. Indentation curve (a) was collected at 10 μm/s and curve (b) was collected at 0.063 μm/s. (c) Rate dependence of the indent curve slope for various humidity values. (Reproduced with permission from reference 23. Copyright 2003 Wiley.)

hydrogels at low humidity are similar to those measured from bulk-dehydrated hydrogels. As most contact lens wearers know from experience, bulk-dehydrated pHEMA hydrogels are glassy. This indicates that at ambient humidity values, the hydrogel/air interface region of the bulk-hydrated hydrogel is dry. At higher humidity values, the rate dependence of the indentation curves becomes more pronounced, as the rate of dehydration from the surface decreases and the surface region of the bulk-hydrated lens contains more water. However, the SFG spectrum in Figure 5 suggests that at the monolayer level, the pHEMA surface is not fully hydrated, even under saturated humidity conditions. SFG spectra suggest that, at the monolayer level, the surface is fully hydrated only when it is in direct contact with water (at the hydrogel/water interface).

Similar data was obtained from the surface of pHEMA+MA.[24] The surfaces of both pHEMA and pHEMA+MA hydrogels are dry and glassy at low humidity and become soft as the humidity increases. However, for intermediate humidity values the surface of pHEMA+MA was measured to be stiffer than the surface of the pHEMA lens. The pHEMA lens surface is softer and retains more water than the pHEMA+MA lens, even though the bulk contains less water. This suggests that the water in the interfacial region of the neutral hydrogel contains more strongly bound water than the ionic hydrogel does. This observation correlates with clinical trials suggesting that ionic hydrogels tend to dehydrate faster on the eye than neutral hydrogels.

Conclusions

SFG spectroscopy experiments are uniquely sensitive to the molecular structures of interfaces and are capable of observing restructuring that occurs as a result of changes in the environment. When information obtained from AFM is combined with that obtained from SFG surface spectroscopy, a more detailed understanding of events occurring at interfaces is obtained. The combination of SFG and AFM was used to study surface segregation and wetting behavior of polyolefin blends. In that example, SFG provided an average surface composition while AFM images were able to show the lateral structure of the surface. SFG was used to probe the surface molecular structure and AFM was used to measure the mechanical properties of hydrogel contact lens materials as a function of hydration. It was shown that the surfaces of pHEMA based contact lenses are likely to be dry under normal wearing conditions.

Acknowledgement

This work was supported by the Director, Office of Science, Office of Basic Energy Sciences, Division of Materials Sciences and Engineering, of the U.S. Department of Energy under Contract No. DE-AC03-76SF00098. Additional support and materials were provided by Dr. Roger Phillips at Basell Polyolefins Inc., and also by Ocular Sciences Inc..

References

1. Shen, Y. R., *Principles of Nonlinear Optics*, John Wiley & Sons, New York, 1984.
2. Wei, X.; Hong, S. C.; Zhuang, X. W.; Goto, T.; Shen, Y. R. *Phys. Rev. E* **2000**, *62*, 5160.
3. Miranda, P. B.; Shen, Y. R. *J. Phys. Chem. B* **1999**, *103*, 3292.
4 Zhuang, X.; Miranda, P. B.; Kim, D.; Shen, Y. R. *Phys. Rev. B* **1999**, *59*, 12632.
5. Hirose, C.; Akamatsu, N.; Domen, K. *Appl. Spect.* **1992**, *46*, 1051.
6. Chen, Z.; Shen, Y. R.; and Somorjai, G. A. *Ann. Rev. Phys. Chem.* **2002**, *53*, 437.
7. Opdahl, A.; Phillips, R. A.; Somorjai, G. A. *J. Phys. Chem. B.* **2002**, *106*, 5212.
8. Opdahl, A; Phillips, R. A.; Somorjai, G. A. *J. Poly. Sci. B.* **2004**, *42*, 421.
9. Kim, S. H.; Opdahl, A.; Marmo, C.; Somorjai, G. A. *Biomaterials* **2002**, 23, 1657.
10. Opdahl, A.; Phillips, R.A.; Somorjai, G. A. *Macromolecules*, **2002**, *35*, 4387.
11. Chen, Z.; Ward, R.; Tian, Y.; Eppler, A. S.; Shen, Y. R.; Somorjai, G. A.; *J. Phys. Chem. B.*, **1999**, *103*, 2935.
12. Wang, H.; Composto, R. J. *Macromolecules,* **2002**, *35,* 2799.
13. Zhang, D.; Shen, Y. R.; Somorjai, G. A. *Chem. Phys. Lett.* **2002**, *281*, 394.
14. Chen, Z.; Ward, R.; Tian, Y.; Baldelli, S.; Opdahl, A.; Shen, Y. R.; Somorjai, G. A. *J. Am. Chem. Soc.* **2000**, *122*, 10615.
15. Zhang, D.; Dougal, S. M.; Yeganeh, M. S. *Langmuir* **2000**, *16*, 4528.
16. Gautam, K. S.; Schwab, A. D.; Dhinojwala, A.; Zhang, D.; Dougal, S. M.; Yeganeh, M. S. *Phys. Rev. Lett.* **2000**, *85*, 3854.
17. Briggman, K. A.; Stephenson, J. C.; Wallace, W. E.; Richter, L. J. *J. Phys. Chem. B* **2001**, *105*, 2785.
18. Oh-e, M.; Hong, S. C.; Shen, Y. R. *Appl. Phys. Lett.* **2002**, *80*, 784.

19. Wang, J.; Chen, C.; Buck, S. M.; Chen, Z. *J. Phys. Chem. B.* **2001**, *105*, 12118.
20. McConville, P.; Pope, J. M. *Polymer*, **2001**, *42*, 3559.
21. Lopez-Alemany, A.; Compan, V.; and Refojo, M. F. *J. Biomed. Mat. Res.*, **2002**, *63A*, 319.
22. Pritchard, N.; and Fonn, D. *Opthal. Physiol. Opt.*, **1995**, *15*, 281.
23. Opdahl, A.; Kim, S. H.; Koffas, T. S.; Marmo, C.; Somorjai, G. A. *J. Biomed. Mat. Res.* **2003**, *67A*, 350.
24. Koffas, T. S.; Opdahl, A.; Marmo, C.; Somorjai, G. A. *Langmuir.* **2003**, *19*, 3453.
25. Wang, J.; Paszti, Z.; Even, M. A.; Chen, Z. *J. Am. Chem. Soc.* **2002**, *124*, 7016.
26. Opdahl, A.; Somorjai, G. A. *Langmuir*, **2002**, *18*, 9409.
27. Sinniah, S. K.; Steel, A. B.; Miller, C. J.; Reutt-Robey, J. E. *J. Am. Chem. Soc.* **1996**, *118*, 8925.
28. Kim, J.; Opdahl, A.; Chou, K. C.; Somorjai, G. A. *Langmuir* **2003**, *19*, 9551.
29. Tanford, C. *The Hydrophobic Effect: Formation of Micelles and Biological Membranes; 2nd* ed.; Krieger: Malabar, FL, 1991.
30. Zhang, X.; Zhu, Y.; Granick, S. *J. Am. Chem. Soc.* **2001**, *123*, 6736.
31. Eisenberg, D.; McLachlan, A. D. *Nature* **1986**, *319*, 199.
32. Muller, N. *Acc. Chem. Res.* **1990**, *23*, 23.
33. *Proteins at Interfaces II: Fundamentals and Applications*; Horbett, T. A.; Brash, J. L., Eds.; ACS: Washington, DC, 1995.
34. Fontyn, M.; van't Riet, K.; Bijsterbosch, B. H. *Colloids Surf.* **1991**, *54*, 349.
35. Chen, C.; Even, M. A.; Wang, J.; Chen, Z. *Macromolecules* **2002**, *35*, 9130.
36. Iler, R. K. *The Chemistry of Silica*; Wiley: New York, 1979.
37. Dreesen, L.; Humbert, C.; Hollander, P.; Mani, A. A.; Ataka, K.; Thiry, P. A.; Peremans, A. *Chem. Phys. Lett.* **2001**, *333*, 327.
38. Stanners, C. D.; Du, Q.; Chin, R. P.; Cremer, P.; Somorjai, G. A.; Shen, Y. R. *Chem. Phys. Lett.* **1995**, *232*, 407.

Chapter 10

Multilayered Nanoscale Systems and Atomic Force Microscopy Mechanical Measurements

H. Shulha, Y.-H. Lin, and V. V. Tsukruk*

[1]Department of Materials Science and Engineering, Iowa State University, Ames, IA 50011
***Corresponding author: vladimir@iastate.edu**

The atomic force microscopy approach developed for the microindentation of layered elastic solids was adapted to analyze nanoprobing of ultrathin (1-10 nm thick) polymer films on a solid substrate. This "graded" approach offered a transparent consideration of the gradient of the mechanical properties between layers with different elastic properties. Some examples of recent applications of this model to nanoscale polymer layers were presented. We considered polymer layers with elastic moduli ranging from 0.1 to 20 MPa in a solvated state. A complex shape of corresponding loading curves and elastic modulus depth profiles obtained from experimental data were fitted by the graded functions suggested.

Introduction

The ability to probe surface mechanical properties with nanometer-scale lateral and vertical resolutions is critical for many emerging applications involving nanoscale (1-10 nm) compliant coatings for microelectromechanical and microfluidic devices where nanoscale details of surface deformations and shearing play a critical role in overall performance.[1,2,3,4,5,6] Usually, a nanomechanical probing experiments exploits either atomic force microscopy (AFM) or microindentation techniques.[7,8] A number of successful applications have recently been demonstrated including nanomechanical probing of spin-coated and cast polymer films, organic lubricants, self-assembled monolayers, polymer brushes, biological tissues, and individual tethered macromolecules. [9,10,11,12,13,14,15,16,17] The absolute values of the elastic modulus have been measured for various polymer surfaces in the range from 0.1 MPa to 20 GPa.

Spatial (concurrent vertical and lateral) resolution on a nanometer scale unachievable by any other probing technique makes AFM nanomechanical probing a unique experimental tool. We believe that a further expansion of the AFM-based probing of ultrathin (below 10 nm) polymer films in a contact mode regime will rely on solving several fundamental issues including the evaluation of substrate effect and the elastic response for multilayered coatings. In this communication, we present our work to adapt known approaches developed for the analysis of microindentation experiments of the elastic layered solids to AFM experiments. The results of corresponding AFM probing of several types of nanoscale polymeric coatings on solid substrates representing two-layered structures are reported and discussed.

A. General contact mechanics background

General relationships between the normal load and the elastic indentation are suggested in classical Hertzian and Sneddon theories, with more complex cases analyzed with the Johnson-Kendal-Roberts approach. [18,19,20,21,22] As considered, indentation depth is a function of the applied force (normal load) P, tip geometry (radius R or parabolic focus distance c), as well as the mechanical and the adhesion properties of the contacting bodies (Figure 1). The normal load for AFM nanomechanical probing experiments conducted is calculated as $P = k_n \cdot z_{defl}$, where k_n is the vertical spring constant of the cantilever deflected in vertical direction by z_{defl}. The above mentioned R (or c) and k_n as well as the Poisson's ratio, ν, are considered to be initial system parameters, which must be known, measured, or calibrated before the nanomechanical analysis.

The most general relationship between indentation depth, h, and normal load, P, in the course of indentation experiment can be presented in the very general form as: [18-22]

$$P = a\,h^{\,b} \qquad (1)$$

where a and b are specific, model-dependent geometrical parameters (e.g., $b = 3/2$ for both Hertzian and parabolic Sneddon's contacts).

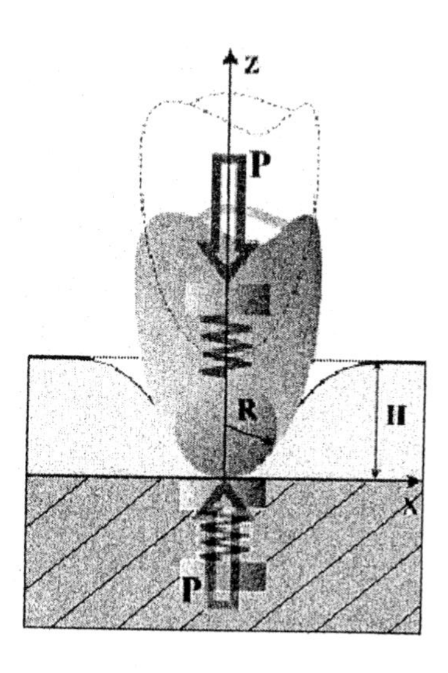

Figure 1. A two-spring model for the analysis of the loading curve for the parabolic tip – plane surface contact.

On the other hand, the Sneddon's model suggests a specific and practical analytical relationship between the surface stiffness, dP/dh and Young's modulus, E', in the form:

$$\frac{dP}{dh} = \frac{2\sqrt{A}}{\sqrt{\pi}} E' \qquad (2)$$

where E' is the composite modulus.

From dP/dh dependence obtained from microindentation experiment and the calculated/measured contact area variation for a specific shape of the indentor (specific analytical expressions are suggested for circular, pyramidal, and parabolic shapes), one can evaluate an absolute value of the elastic modulus. For a routine estimation of the elastic modulus value for small indentation depths, the Hertzian model of a sphere-plane contact type is applied.

B. Mechanical models for multilayered elastic solids

There are several methods for the evaluation of the elastic modulus of thin films on solid substrates, which include microindentation and bending experiments.[23,24,25,26,27,28,29,30,31,32] All of them have own ranges of applicability and limitations. A model, which considered the elastic deformation of the layered solids with a certain transfer of the mechanical load between adjacent layers, was proposed for the analysis of microindentation data.[33,34] A general key point of this approach was the suggestion to represent the composite compliance of two-layered solids (e.g., such as a film-substrate system) as a superposition of individual compliances in the general form:

$$\frac{1}{E'} = \frac{1}{E_f} \cdot \left(1 - e^{-\alpha \cdot h/t}\right) + \frac{1}{E_s} \cdot \left(e^{-\alpha \cdot h/t}\right) \tag{3}$$

where E_f, E_s are elastic moduli of the film and the substrate, t is the total thickness of the film, h is the indentation depth, and α is a parameter defining contributions of different layers. This approach introduces a new measure of a level of the transfer of the mechanical deformation between layers represented by a specially selected function, the transfer function, $e^{-\alpha h/t}$. This transfer function depends upon the total thickness of the layer, indentation depth, and the properties of the inter-layer interactions as reflected by the parameter α. The transfer function for the elastic layered solid has an initial small value for very small, initial deformations ($h << t$) and increases for larger deformations ($h \leq t$).

Below, we will show that a generalized approach that starts with a simple definition of the depth profile as a smooth function with gradual localized changes provides a means for the "visualization" of the transfer function and its concise interpretation for complex layered solids with two- and tri-layer architectures.

Model for Layerd Systems

A. Graded functions

Here we represent a model developed for multilayered system. Details and derivation can be found elsewhere.[35] The model proposed here is based on simple initial assumptions on the gradient properties of the layered systems and very basic arguments. It combines all major features suggested separately in several approaches mentioned above, and leads to a relatively simple equation for the description of the elastic modulus gradient for two-layered systems in the form:

$$E' = \frac{E_1 - E_0}{1 + \exp(-\alpha(E_1 - E_0)(h - h_0)/(E_0 h_0))} \tag{4}$$

where E_o and E_1 are two levels of the elastic modulus for a two-layered system, h_o is a point where the modulus equals half of the difference between the two layers, $E'(h_0) = (E_0 - E_1)/2$, and α is a coefficient of proportionality.

Overall, the depth profile of the elastic modulus for a two-layered system can be described as a superposition of the initial level and variable contribution for the two-layered system (4) as:

$$E(h) = E_o + E'(h) \tag{5}$$

Modulus profile in this form can be directly used in equation (2) for fitting experimental data for the elastic response at a variable indentation depth.

For the important case of a compliant film on a stiff substrate, only two unknown variables can be varied to fit experimental data (the elastic modulus of the top layer and parameter α) assuming that the elastic properties of the substrate are known. Usage of the proposed approach allows for the analysis of different layered structures. General profile of elastic modulus distribution as described by equation (4) is presented in Figure 2. Two different levels of the elastic modulus are separated by a transition zone with a gradient of the elastic properties.

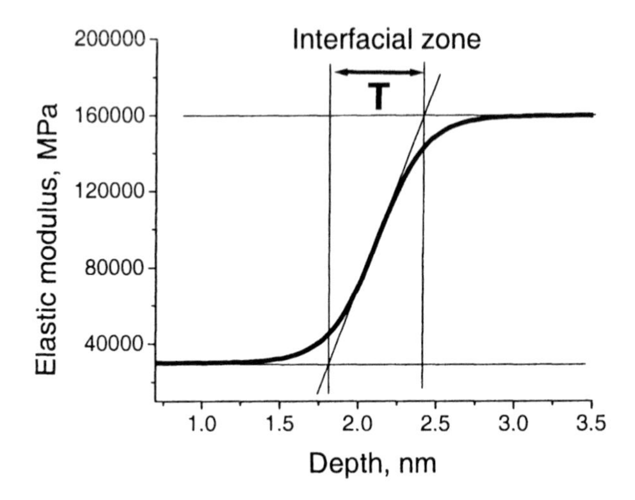

Figure 2. The interfacial zone, T, is defined for this profile as discussed in the text.

The variation of the elastic modulus in a wide range creates very different profiles including a virtually uniform distribution for a layered system with a small difference in elastic moduli (Figure 3). It is worth noting that here and below in Figures 3-5, the absolute values of the elastic modulus and the indentation depth were selected only for illustrative purposes, can be very different for different layered models, and are not to be used for the comparison between different layered models.

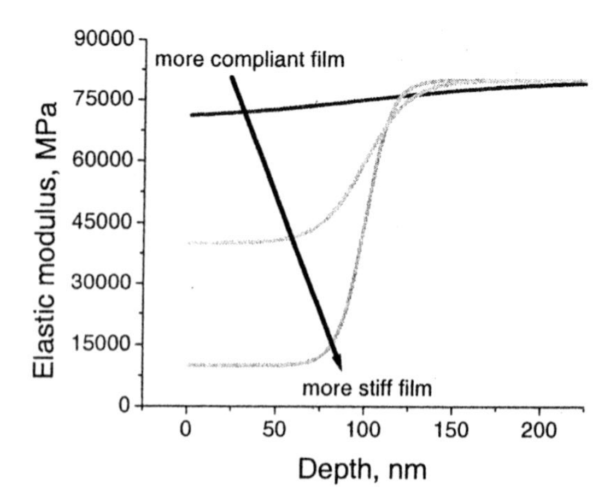

Figure 3. Influence of the elastic modulus of the topmost layer on the total depth distribution of the elastic modulus of the two-layered solid (E_1 = 80000MPa).

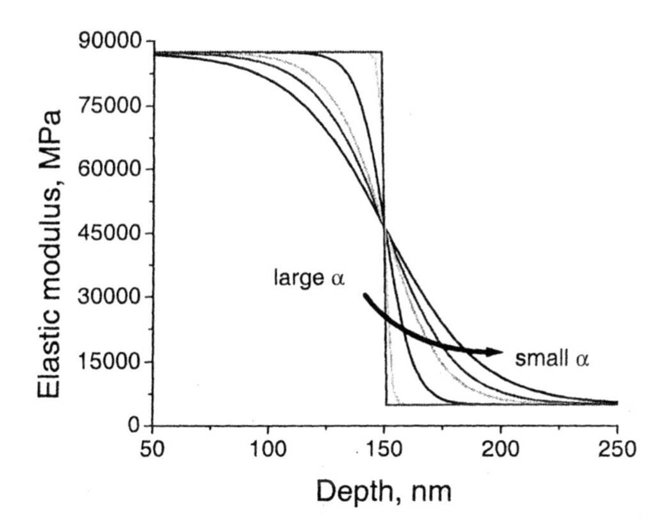

Figure 4. Influence of the transition parameter α on the overall shape of a hypothetical two-level depth profile of the selected elastic moduli (E_o = 85000MPa and E_1 = 5000MPa).

The width of this transition zone between two layers is determined by the parameter α through the transition function. A high value of this parameter corresponds to a very sharp interfacial zone resulting in a step-like shape of $E'(h)$ (Figure 4). Decreasing α-value results in a gradual broadening of the step-function with the formation of a virtually continuous gradient for a very low value of α (Figure 4).

For practical purposes, instead of "non-transparent" parameters α, we introduce the thickness of the transition zone as the major fitting parameter as illustrated in Figure 2. We calculate the "effective" thickness of the transition zone, T, as a distance between two points representing the intersections between two levels of elastic moduli and the slope to the fitting curve in the point where the value equals the average of the above mentioned moduli. This parameter has direct physical meaning as a measure of the sharpness of the transition zone between layers (see Figure 2 for the definition of the thickness of the transition zone). A reasonable physical value of this parameter and its correlation with expected or known structural gradients (e.g., controlled by processing, deposition, or synthetic routine) are important verifications of the concept and the fitting procedure as will be discussed below.

Using this expression, even more complicated a tri-step graded function can be simulated to analyze the surface structure with a complex profile of stiff on compliant on stiff type (Figure 5). By the variation of the transition zone gradient, this step-function can be converted from a sharp step function to a smooth function with a minor depletion in the middle. On the other hand, by changing the level of the elastic modulus for the intermediate layer, this function can be converted to the tri-layer function with ascending or descending elasticity (Figures 5).

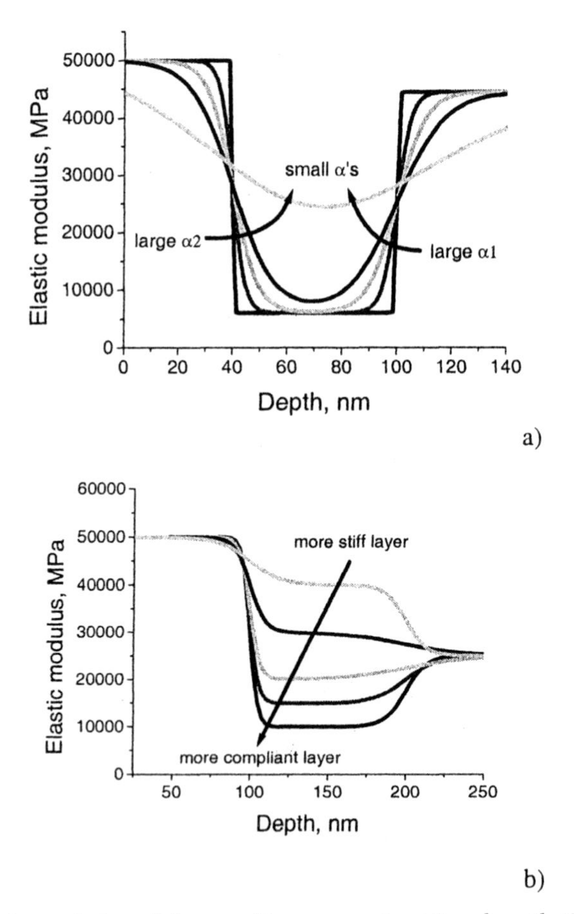

Figure 5. a) The variation of the transition parameter α in a hypothetical tri-layered system with selected values of elastic moduli (E_o = 50000MPa, E_1 = 5000MPa and E_2 = 45000MPa). Decreasing α results in the broadening of the transition zone and disappearance of the sharp steps. b) Influence of the elastic modulus of the interlayer on the depth distribution of the elastic modulus of the tri-layer system presented above (E_o = 50000MPa and E_2 = 25000MPa).

B. Issues with nanomechanical probing.

Here, it is important to mention several other critical contributions, which could affect force-distance data and are not accounted for in the model discussed here. First, strong adhesion between the AFM tip and surface can disturb the initial portion of loading curves and result in significant overestimation of the elastic modulus level at small indentation depth. We analyzed this contribution in our previous publication [36] and demonstrated significance of the adhesion hysteresis and initial non-zero contact area for very compliant polymeric materials with high adhesion (e.g., polar rubbery layers with the elastic modulus below 2 MPa and surface energy much higher than 20 mJ/m^2) in air. For these materials, an application of the Hertzian model resulted in manifold overestimation of the elastic modulus for small indentations. A complete Johnson-Roberts-Kendall model should be applied that makes consideration significantly more complex and requires additional non-trivial measurements. However, for compliant materials with modest adhesion and higher stiffness, this overestimation is limited to a few initial several data points and, thus, the approach discussed here can be applied.

Second, a viscous contribution (time-dependent mechanical properties) can be critical in defining an overall shape of loading curves for viscoelastic polymeric materials. As we discussed earlier, this phenomenon would result in a concave shape of the force-distance curves, which, in fact, is sometimes observed.[27,40] This contribution can be treated by applying Johnson's recent development [37] as was discussed in separate publications. [27,38] Third, the surface roughness of the layers studied here is extremely low, thus, virtually eliminating any significant scattering in the first few data points observed for rough surfaces.

Experimental Data for the Selected Layered System: Polymer Brushes in Fluid

Here, we present a recent example of application of the approach described here to a complex case of a very complaint polymer layer grafted to a solid substrate and placed in selective solvent. AFM experiments were done according to the experimental procedures described elsewhere.[39,40] Spring constants were selected in the range from 0.1 to 0.4 N/m depending on sample elasticity and measured according to described technique.[41] Tip radius was measured with a gold nanoparticle reference standard.[42] Experimental force-distance data (cantilever deflection versus piezoelement displacement) collected in the force-volume mode were processed by *MManalysis* software package developed in our laboratory with an added option for the multilayered analysis. [43]

Polymer brush layers are composed of macromolecules chemically grafted to a solid silicon substrate through a mediating functionalized self-assembled monolayer (Figure 6).[44] In the case considered here, a binary polymer brush layer was prepared from rubbery poly(acrylic acid.) (PAA) and glassy poly(styrene) (PS), both with high molecular weight, grafted to a silicon wafer by the "grafting to" approach as described elsewhere.[45,46,47] An array of force-distance curves was collected for the surface areas of 0.7 x 0.7 μm. Surface topographical images were obtained before and after the experiments to confirm that deformations were elastic.

An example of the force-distance curve for the binary polymer brush (PS/PAA) placed in a good solvent for PAA (water) is presented in Figure 7. Under this solvent, polymer chains were highly swollen and possessed very low elastic modulus as expected for polymers in a good solvent.[48] Under these conditions, force-volume probing with very low normal forces generated an array of force-distance curves with a complex shape showing two regions with different slopes.

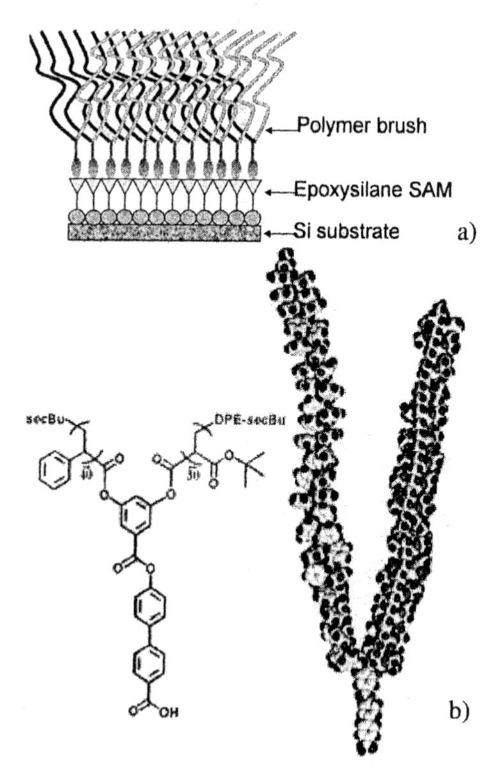

Figure 6. a) A general sketch of Y-shaped polymer brush layer with polymer chains tethered to the solid substrate via SAM.[45] b) Chemical formula and molecular model of PS/PAA brush.[47]

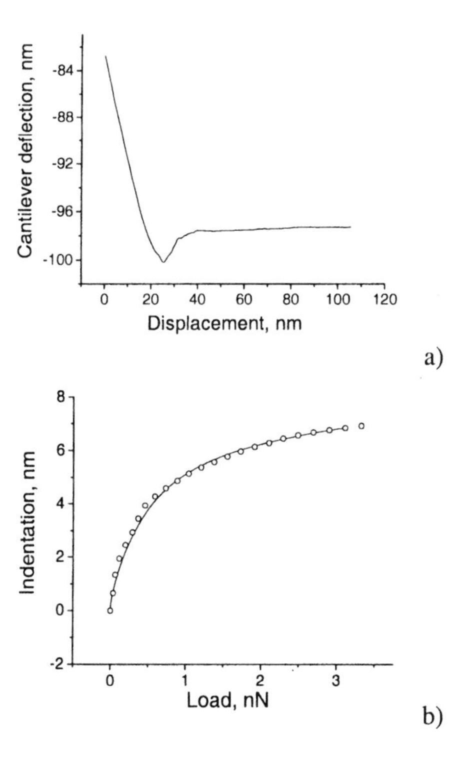

Figure 7. a) An experimental average loading force distance curve obtained for the system under water on stiffer surface in Y-shaped binary brush layer.[46,47] *b) Fitting the experimental loading curve (circles) by the two-layered model (solid line).*

Accordingly, the loading curve with the indentation depth reaching 8 nm displayed a complex shape, which deviated significantly from normal Hertzian behavior expected for a uniform elastic material (Figure 7). Individual force-distance curves had similar shapes with higher level of noise removed by the averaging of a significant number of experimental curves.

The best fitting of the loading curve and the depth profile can be obtained by using a very low elastic modulus value of 9 MPa for initial deformations not exceeding 6 nm. This very compliant region is replaced with rising elastic resistance for larger indentation depth caused by the presence of the underlying solid substrate (Figure 8). But it is worth to note that the stable compression of the hard solid substrate can not be done with this soft cantilever according to the previous estimates.[49]. That is why the experiment was conducted on the depth that did not exceed the height of the molecular in the swollen state and the brush was considered as two-layered system.

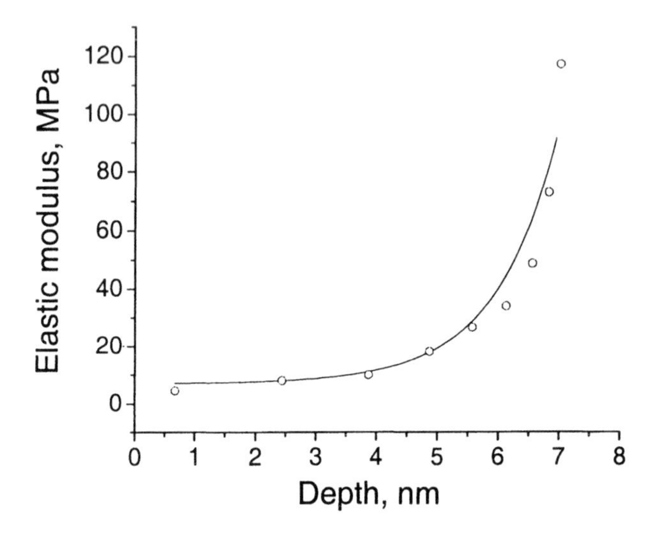

Figure 8. Fitting the experimental data on the depth distribution of the elastic moduli of polymer layer in water (circles) with the two-layered model (solid line).

Conclusions

We suggest that the approach proposed, if applied carefully and properly, can be used for the analysis of multilayered elastomeric materials with an appropriate thickness of different layers and efficient load transfer between layers and between layers and substrates under conditions of purely elastic deformation. The approach has limitations related to its eligibility only for purely elastic, completely reversible deformations without any contribution from plastic deformation, viscoelastic phenomenon, strong adhesion, and high friction. We demonstrate that despite these limitations, the approach proposed can be successfully applied toward complex surface layers including polymer layer swollen in solvents with potentially complex distribution of the elastic properties.

Acknowledgements

We acknowledge useful discussions with Dr. V. V. Gorbunov. This work is supported by The National Science Foundation, CMS-0099868 and DMR-0308982 Grants.

References

1. Muller, R. S. in *Micro/Nanotribology and Its Applications*, edited by B. Bhushan (Kluwer Press, **1997**), p. 579.
2 Mate, C. M., and Wu, J., in *Microstructure and Microtribology of Polymer Surfaces*, edited by Tsukruk, V. V., and Wahl, K., (ACS Symposium Series **2000**, *741*), p. 405.
3 *Micro/Nanotribology and Its Applications*, edited by Bhushan, B., (Kluwer Press, **1997**).
4 *Tribology Issues and Opportunities in MEMS*, edited by Bhushan, B., (Kluwer Academic Press, **1998**).
5 Komvopoulos, K., *Wear* **1996,** *200,* 305.
6 Delamarche, E.; Bernard, A.; Schmid, H.; Michel, B.; and Biebuyck, H., *Science*, **1997,** *276,* 779.
7 Chen, X.; and Vlassak, J., *J. Mater. Res.,* **2001,** *16,* 2974.
8 Tsukruk, V. V., *Adv. Materials,* **2001,** *13,* 95.
9 Matthews, A.; Jones, R.; and Dowey, S., *Tribolgy Lett,* **2001,** *11,* 103.
10 Vanlandingham, M. R.;.McKnight, S. H.; Palmese, G. R.; Ellings, J. R.; Huang, X.; Bogetti, T. A.; Eduljee, R. F.; and Gillespie, J. W., *J. Adhesion,* **1997,** *64,* 31.
11 Bliznyuk, V. N.; Everson, M. P.; and Tsukruk, V. V., *J. Tribology,* **1998,** *120,* 489.
12 Kurokawa, T.; Gong, J. P.; and Osada, Y., *Macromolecules,* **2002,** *35,* 8161.
13 Domke, R.; and Radmacher, M., *Langmuir,* **1998,** *14,* 3320.
14 Lemoine, P.; and Mc Laughlin, J., *Thin Solid Films,* **1999,** *339,* 258.
15 Eaton, P.; Fernandez, E. F.; Ewen, R. J.; Nevell, T. G.; Smith, J. R.; and Tsibouklis, J., *Langmuir,* **2002,** *18,* 10011.
16 Gorbunov, V.; Fuchigami, N.; Stone, M.; Grace, M.; and Tsukruk, V. V., *Biomacromolecules,* **2002,** *3,* 106.
17 Tsukruk, V. V.; Shulha, H.; and Zhai, X., *Appl. Phys. Lett,* **2003,** *82,* 907.
18 Oliver, W.; and Pharr, G., *J. Mater. Res.,* **1992,** *7,* 1564.
19 Sneddon, I. N., *Int. J. Engng. Sci.,* **1965,** *3,* 47.
20 Johnson, K. L.; Kendall, K.; and Roberts, A. D., *Proc. R. Soc.,* London **1971,** *A324,* 301.
21 Pharr, G. M.; Oliver, W. C.; and Brotzen, F. B., *J. Mater. Res.,* **1992,** *7,* 613.
22 Johnson, K. L., Contact Mechanics (Cambridge university press, Cambridge, **1985**).

23 Nix, W. D., *Metal. Trans.*, **1989,** *20A*, 2217.

24 Pharr, G. M.; and Oliver, W. C., *MRS Bull.*, **1992,** *17,* 28.

25 Field, J. S.; and Swain, M. V., *J. Mater. Res.*, **1993,** *8,* 297.

26 Makushkin, A. P., *Friction and Wear,* **1990,** *11(3),* 423.

27 Chizhik, S. A.; Gorbunov, V. V.; Luzinov, I.; Fuchigami, N; and Tsukruk, V. V., *Macromol. Symp.*, **2001,** *167,* 167.

28 Gao, H.; Chiu, C. H.; and Lee, J., *Int. J. Solids Structures,* **1992,** *29,* 2471.

29 Pender, D.; Thompson, S.; Padture, N.; Giannakopoulos, A.; and Suresh, S., *Acta Mater.* , **2001,** *49,* 3263.

30 Suresh, S., *Science,* **2001,** *292,* 2447.

31 Giannakopoulos, A. E.; and Suresh, S., *Int. J. Solid Structures,* **1997,** *34,* 2357.

32 Giannakopoulos, A. E.; and Suresh, S., *Int. J. Solid Structures,* **1997,** *34,* 2393.

33 Doerner, M. F.; and Nix, W. D., *J. Mater. Res.*, **1986,** *1,* 601.

34 King, R. B., *Int. J. Solids Structures,* **1987,** *23,* 1657.

35 Kovalev, A.; Shulha, H.; Lemieux, M.; Myshkin, N.; and Tsukruk, V. V., *J. Mat. Res.*, **2004,** *19,* 716.

36 Chizhik, S. A.; Huang, Z.; Gorbunov, V. V.; Myshkin, N. K.; and Tsukruk, V. V., *Langmuir,* **1998,** *14,* 2606.

37 Johnson, K. L., in *Microstructure and Microtribology of Polymer Surfaces,* edited by Tsukruk, V. V. and Wahl, K. (ACS Symposium Series **1998,** *741,* USA), p. 24.

38 Shulha, H.; Kovalev, A.; Myshkin, N.; and Tsukruk, V. V., *Europ. Polym. J.*, **2004,** *40,* 949.

39 Shulha, H.; Zhai, X.; and Tsukruk, V. V., *Macromolecules,* **2003,** *36,* 2825.

40 Tsukruk, V. V.; and Gorbunov, V. V., *Probe Microscopy,* **2002,** *3-4,* 241.

41 Hazel, J.; and Tsukruk, V. V., *Thin Solid Films,* **1999,** *339,* 249.

42 Radmacher, M.; Tillmann, R.; and Gaub, H., *Biophys. J.*, **1993,** *64,* 735.

43 Tsukruk, V. V.; and Gorbunov, V. V., *Microscopy Today,* **2001,** *01-1,* 8.

44 Luzinov, I.; Julthongpiput, D.; Liebmann-Vinson, A.; Cregger, T.; Foster, M. D.; and Tsukruk, V. V, *Langmuir,* **2000,** *16,* 504.

45 Lemieux, M.; Minko, S.; Usov, D.; Stamm, M.; and Tsukruk, V. V., *Langmuir,* **2003,** *19,* 6126.

46 Julthongpiput, D.; Lin, Y.-H.; Teng, J.; Zubarev, E. R.; and Tsukruk, V. V., *Langmuir,* **2003,** *19,* 7832.

47 Julthongpiput, D.; Lin, Y.-H.; Teng, J.; Zubarev, E. R.; and Tsukruk, V. V., *J. Am. Chem. Soc.*, **2003,** *125,* 15912.

48 Lemieux, M.; Usov, D.; Minko, S.; Stamm, M.; Shulha, H.; and Tsukruk, V. V., *Macromolecules,* **2003,** *36,* 7244.

49 Huang, Z.; Chizhik, S. A.; Gorbunov, V. V.; Myshkin, N. K.; Tsukruk, V. V.. *ACS Symposium Series,* **2000,** *741,* 177.

Single-Molecule Studies

Chapter 11

Probing Single Polymer Chain Mechanics Using Atomic Force Microscopy

Pamela Y. Meadows, Jason E. Bemis, Sabah Al-Maawali, and Gilbert C. Walker

Department of Chemistry, University of Pittsburgh, Pittsburgh, PA 15260

Abstract

Polymer and protein adsorption play important roles in processes such as the fouling of ships to determining the biocompatibility of surfaces. By understanding the mechanics involved in polymer-surface interactions, new insight can be obtained into designing surfaces that resist biofouling. With the atomic force microscope's (AFM) unique ability to measure small magnitudes of force (< 20pN) as a function of tip-sample separation, it has become an essential tool for studying ligand-receptor adhesion forces, polymer elasticity, and the folding kinetics of many biological systems. In the review presented here, we will investigate the elastic properties of polystyrene-b-poly-2-vinylpyridine (PS-P2VP) chains from spun-cast films,[1] characterize the polydispersity of poly(dimethylsiloxane) (PDMS) surfaces,[2] and estimate the stability of an adhesion promoting protein (fibronectin, FN) as a function of protein surface density.[3]

Introduction

Single Polymer Chain Elongation by Atomic Force Microscopy[1]

Studies of PS-P2VP adhesion to an AFM probe were examined and modeled. Adhesion models between two surfaces such as Johnson, Kendall, and Roberts theory (JKR)[4] could not be used since the macroscopic models fail to describe the adhesion of surfaces linked by a single or multiple polymer chains.

Figure 1 shows a typical force plot obtained for PS60800-P2VP46900 (the numbers after the block name indicate the molecular weight of each individual block in units of Daltons). As the tip contacts the surface, one or more polymer chains can physisorb to the tip. Once the AFM piezo retracts from the surface (A), the polymer chains become elongated as can be seen in B and C of Figure 1. Here, the AFM probe experiences elastic tension due to the attached polymer chain, giving rise to the increase in force. At point C, the polymer chain breaks free from the tip, resulting in a return to zero tip deflection (D). The length (58 nm) and force (180 pN) at point C characterize the behavior of the extended polymer chain; data between points A through C are fit by theoretical models (dotted and solid lines).

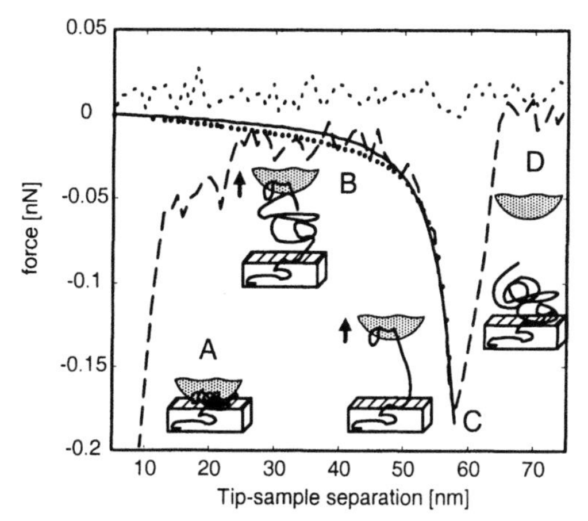

Figure 1: A force plot obtained for the extension of PS60800-P2VP46900 with tip illustration of polymer dynamics occurring during forced extension. The solid line is a WLC fit to the data giving parameters of 5.75 Å (persistence length), 0.90 (extension ratio), and a χ^2 of 1.52. The dotted line represents a FJC fit with 5.86 Å obtained for the Kuhn length, 0.96 for the extension ratio, and a χ^2 of 1.45. (Reproduced with permission from reference 1. Copyright 1999)

Two principal models[5] used to explain the entropic elasticity of polymer stretching events are the freely jointed chain (FJC) and the wormlike chain (WLC) models. In the FJC model, the force-distance dependence is described by Equation 1 where $L^*(R)$ is the inverse Langevin function, $L^*(R) = \beta$, with the Langevin function $L(R) = \coth(\beta) - 1/\beta$.

$$F = \left(\frac{kT}{A_k}\right)L^*(R) \tag{1}$$

Here F represents the tension between two points in units of nN, k is the Boltzmann constant (aJ/K), T is the temperature in units of Kelvin, and A_k is the Kuhn length in units of nm. R represents the unitless extension ratio and can be defined as the fraction of the polymer's contour length that is extended. In the WLC model, the force-distance behavior is described by Equation 2 where q is the persistence length in nm.

$$F = \left(\frac{kT}{q}\right)\left(\frac{1}{4(1-R)^2} - \frac{1}{4} + R\right) \tag{2}$$

Figure 1 demonstrates a fit of the polymer elastic response to both the FJC model (dotted line) and the WLC model (solid line). The major difference in the fitting of the data by these two models occurs in the low force, low extension region (25 to 45 nm). However, due to the noise in the data and as Table 1 demonstrates with the χ^2 values, neither model can be assigned as the better fit. The analysis of the distribution using these models can be seen in Table 1.

Table 1: Fitted Parameters for the WLC and FJC Models

	WLC		**FJC**	
Polymer	median persistence length, Å	mean χ^2	median Kuhn length, Å	mean χ^2
PS7800-PVP10000	3.0 ± 2.6	0.70 ± 0.15	3.6 ± 2.7	0.74 ± 0.17
PS13800-PVP47000	2.4 ± 2.1	0.41 ± 0.26	3.0 ± 4.0	0.49 ± 0.35
PS52400-PVP28100	2.9 ± 11.3	0.40 ± 0.18	4.0 ± 11.6	0.42 ± 0.18
PS60100-PVP46900	4.5 ± 14.0	0.72 ± 0.44	5.8 ± 10.5	0.70 ± 0.44

Evidence of Single Molecule Measurements

It is now important to determine whether the single rupture force plots seen in Figure 1 represent the extension of a single polymer chain or the extension of multiple chains. Using Equation 3, it is possible to estimate the chain diameter that is being elongated with the AFM tip.

$$D = \left(\frac{32qkT}{\pi E} \right)^{\frac{1}{4}} \tag{3}$$

Here D represents the chain diameter, q is the persistence length obtained from a WLC fit, k is the Boltzmann constant, T is temperature, and E represents the Young modulus which has a value of 3 GPa for polystyrene. Analysis shows the molecular diameter to be 2.5 ± 0.5 Å which suggests that either a single chain or chains in series are being extended.

We can eliminate the extension of chains in series as a probable event by looking at the polymer contour length obtained from the AFM measurments. The mean length of polymer extension obtained with AFM is plotted *versus* the chain length (estimated from the covalent radii of carbon, the carbon-carbon bond angle, and the number of monomers in the chain) in Figure 2 (solid line). As the chain length increases, the probability of entanglement increases, and therefore, the length to which the AFM probe can extend the polymer from the surface is smaller as well. Also in Figure 2, an estimate of the contour length of

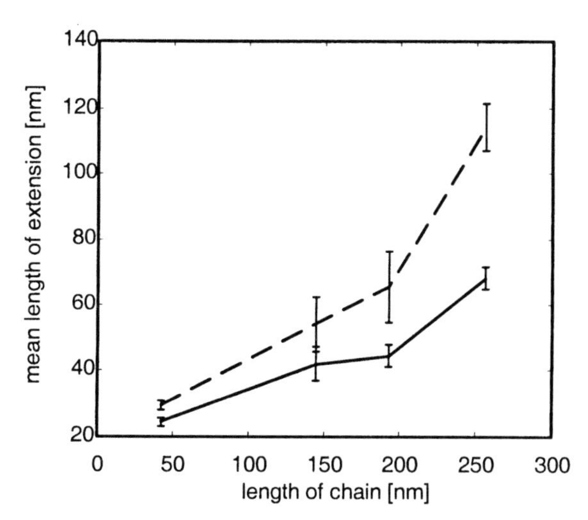

Figure 2: A plot of four block copolymers of PS-P2VP of varying molecular weights. The solid line is a plot of the rupture length obtained from AFM measurements versus the chain length. The upper dashed line is the chain contour length between the tip and surface estimated from the extension ratios from the WLC fit. (Reproduced with permission from reference 1. Copyright 1999)

the chain between the tip and surface is plotted *versus* the chain length (dashed line). The contour length estimation was made using the extension ratio obtained from the WLC fits. Since the estimated contour length is significantly less than the length estimated from the molecular weight, we can exclude the case that chains in series are being extended from the surface by the AFM tip. Further evidence for single molecule measurements comes from the persistence length of 3 to 4 Å. Data obtained from X-ray and neutron scattering have obtained values on this same order,[6] providing further evidence that the single rupture force plots involve the extension of a single polymer chain.

Multiple Elastic Responses Can Indicate Single Molecule Measurements

Figure 3 illustrates a less frequent case where multiple elastic ruptures are observed when the tip retracts from the polymer covered surface. The origin of these ruptures could arise from multiple chains breaking free from the tip or multiple attachments of a single polymer chain rupturing from the AFM probe. To determine whether the multiple ruptures correspond to a single molecule or

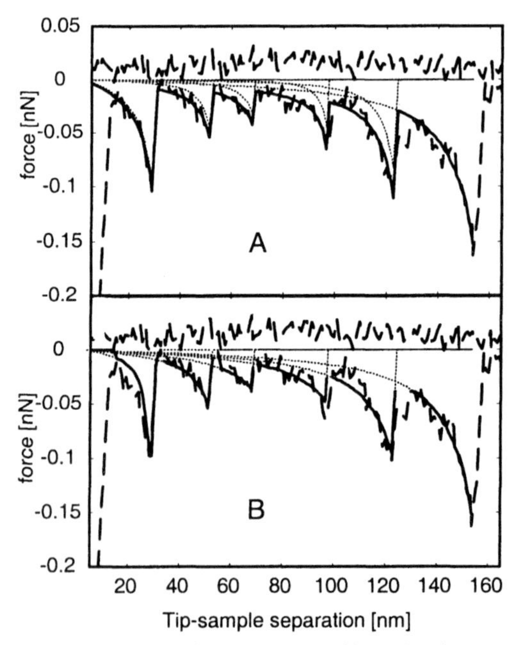

Figure 3: Force plot of PS60800-P2VP46900 displaying multiple elastic responses. A) The multichain, multipersistence length WLC model (solid line) with the dashed lines representing the individual chain responses gives the following values for persistence length (from left to right): 1.9, 12, 22, 63, 21, and 4.7 Å. The χ^2 is 1.09. B) The single chain, single persistence length WLC model gives a persistence length of 0.44nm with a χ^2 of 1.96. (Reproduced with permission from reference 1. Copyright 1999)

multiple polymer chains, the data were fit by several models. In Figure 3A a multichain, multipersistence length model was used while the data in B was modeled by a single chain with a single persistence length. As noted in the caption, the multichain model yielded higher persistence lengths. The data were also fit to a multichain, single persistence length model (not shown), giving a persistence length of 5.7 Å and a χ^2 of 2.54 which is greater than the single molecule, single persistence length model. Because of the correct persistence lengths and molecular diameters obtained with the single chain model compared to the multichain models, it can be concluded that the multiple ruptures observed in Figure 3 are more likely to arise from multiple polymer attachments of a single chain to the AFM tip rather than the extension of multiple polymer chains.

Polydispersity of Grafted Poly(dimethylsiloxane) Surfaces Using Single-Molecule Atomic Force Microscopy[2]

Synthesis of polymers through polycondensation or free radical mechanisms produce molecules of various molecular weights and hence lengths.[7-9] The polydispersity index (PI) is used to describe the chain length distribution of the polymer solution and can be expressed by Equation 4.

$$PI = \overline{M}_w / \overline{M}_n \qquad (4)$$

Here $\overline{M}_w$ is the weight average molecular weight (g/mol) where $\overline{M}_n$ is the number average molecular weight. If the solution is monodisperse, the polydispersity index will have a value of 1. Gel permeation chromatography (GPC), light scattering, and vapor pressure osmometry have been used to quantify the polydispersity of polymer solutions, but few methods exist that quantify the molecular distribution at surfaces.[7,9] Here, we use an AFM to study poly(dimethylsiloxane) (PDMS) and its molecular distribution on hydroxyl terminated silicon surfaces. Molecular weights of 3000 and 15000-20000 were studied with contour lengths of 11 and 50-80 nm respectively.

Figure 4 and Table 2 show the distribution of polymer chain lengths obtained with varying volume ratios of PDMS and CH_2Cl_2 as obtained by AFM extension measurements and the WLC model. Data obtained from GPC can also be seen in Figure 4. The shift of the distribution toward higher contour lengths with increasing peak widths as the PDMS:CH_2Cl_2 volume ratios increase indicates a preferential adsorption of longer chains at higher volume ratios. At the most dilute PDMS:CH_2Cl_2 ratio, the AFM data correlate well with analysis performed using GPC.

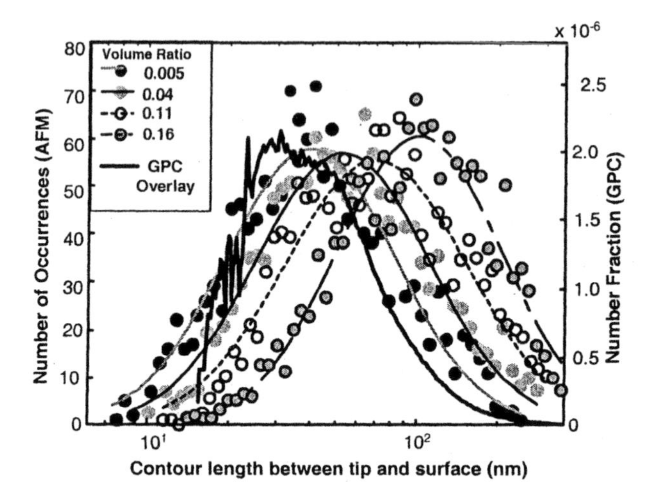

Figure 4: AFM distributions of the contour length obtained for different PDMS:CH_2Cl_2 volume ratios along with the results from GPC. (Reproduced with permission from reference 2. Copyright 2001.)

Table 2: Peak Heights and Widths at Different PDMS:CH_2Cl_2 Volume Ratios using Gaussian fits

PDMS:CH_2Cl_2 Volume Ratios	Peak Positions at Highest Probabilities (nm)	Peak Widths (nm)	PI
0.005	40.0	63.2	1.56
0.04	52.7	85.4	1.56
0.11	69.9	119.7	1.47
0.16	101.0	152.7	1.4

The observance of preferential adsorption of higher molecular weight polymers at higher concentrations is consistent with other studies that analyzed the remaining unreacted solution.[7-9] This observation is illustrated qualitatively in Figure 5 using Flory-Huggins theory[7] where the entropy of mixing is plotted *versus* the polymer chain length. Longer chains are more likely to react at the surface since their entropy of mixing per mass is lower than shorter chains.

From the AFM results for the contour lengths, an estimation of the polymer molecular weight can be made using Equation 5 where the molecular weight of one siloxane monomer is 74 and 0.28 nm is the length of one monomer.

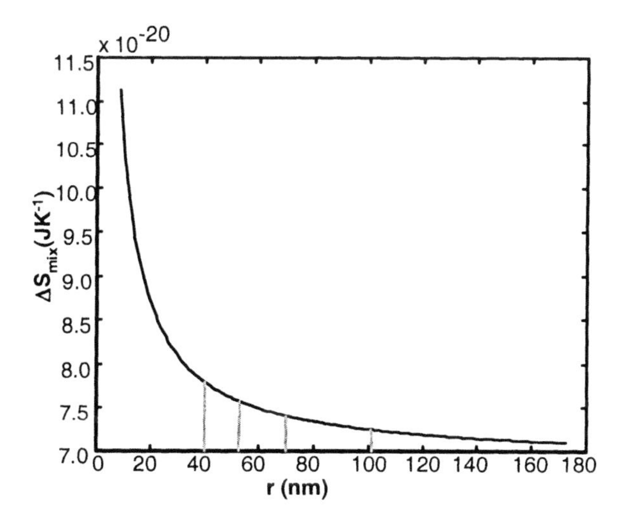

Figure 5: A plot of the entropy of mixing as a function of polymer chain length using Flory-Huggins theory.[7] The lines indicate peak positions of the AFM results for contour lengths of PDMS at different PDMS:CH_2Cl_2 ratios. (Reproduced with permission from reference 2. Copyright 2001.)

$$M_i = \frac{74L_i}{0.28} \tag{5}$$

The polydispersity index can then be calculated using Equation 4 after calculating the weight average molecular weight and number average molecular weight using Equations 6 and 7.

$$\overline{M}_n = \frac{1}{N}\sum M_i \tag{6}$$

$$\overline{M}_w = \frac{\sum_i M_i^2}{\sum_i M_i} \tag{7}$$

The PI obtained from the AFM stretching experiments (1.56) was close to the value obtained from GPC solution measurements (1.62) at lower PDMS:CH_2Cl_2

ratios. Thus, there is significant potential for AFM to analyze contour lengths of macromolecules at surfaces.

Single Molecule Force Spectroscopy of Isolated and Aggregated Fibronectin Proteins on Negatively Charged Surfaces[3]

Much emphasis has been placed on understanding protein adsorption on surfaces. Failure of many implant materials arises because of protein adsorption to surfaces which then results in a cascade of reactions, some of which involve cellular adhesion, activation, and formation of thrombi. Surface chemistry, topography, protein concentration, solvent viscosity, and ionic strength are just a few factors that have been observed to regulate protein and cellular adhesion. In the work presented here, we use an AFM to obtain information about an adhesion promoting protein's stability on negatively charged surfaces in the hope of gaining insight on a surface's effect on protein adsorption.

Mechanical Behavior of Single, Isolated Fibronectin Molecules on a Mica Substrate

Images of the prepared surfaces were collected in intermittent contact mode followed by molecular pulling experiments in *region-specific* surface areas. Figure 6 shows two images obtained in water purified to 18MΩ·cm resistivity, one of a control surface and one with fibronectin (FN) deposited for 5 minutes at a concentration of 1μg/mL. Phosphate buffered saline (PBS), although resembling physiological conditions, could not be used because protein desorption prevailed, resulting in tip contamination.

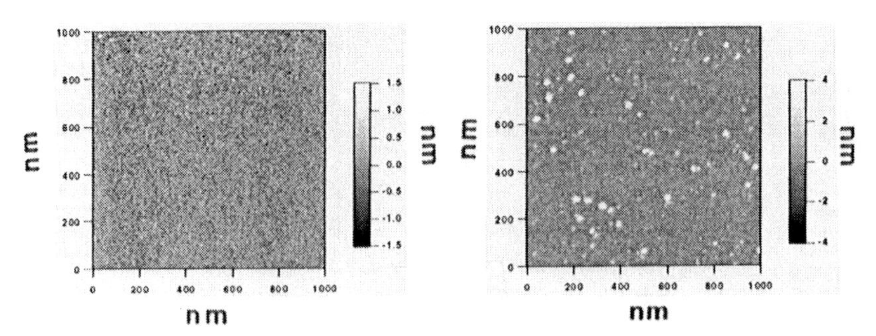

Figure 6: Images showing before (left) and after (right) protein deposition on mica. The larger molecules in the right image represent single molecules of fibronectin. (Reproduced with permission from reference 3. Copyright 2003.)

As can be seen in Figure 6, the proteins adopt a more compact conformation with a height exceeding 8 nm. In a rough approximation, a single FN molecule adopting a spherical conformation should have a height close to 10 nm, indicating the proteins in Figure 6 are not aggregated but rather single FN molecules which were then selected for extension measurements.

To ensure that only single, isolated molecules were probed using uncontaminated tips, we first collected images of the surface and collected our force-extension measurements on *selected* regions of the surface. Force plots on bare regions of the substrate were then obtained with less than 0.1% of the force plots showing stretching events, indicating the ruptures in previous force plots had not been due to a contaminated tip. High-resolution topographic imaging was also performed after pulling experiments, demonstrating the tip had not become contaminated. These images also provided proof that the selected proteins had remained adhered to the substrate during probing.

Figure 7 displays the results from the extension of single, isolated FN molecules on a mica surface. As can be seen in this Figure, the difference in length between successive ruptures was calculated using the points of the cantilever's maximum extension. When neglecting the last rupture, indicative of protein-tip rupture, a most probable length of 9.5 ± 0.5 nm was obtained. This length does not correlate with literature reports (25.1 ± 0.5[10] or 28.5 ± 4.0 nm[11]) for Type III domain unfolding events, a domain type found in fibronectin.

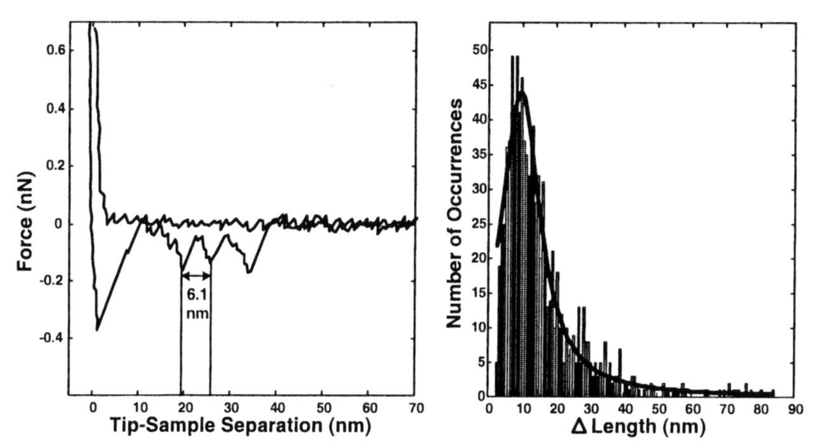

Figure 7: (Left) A force plot obtained for a single, isolated FN molecule on mica. The lines at the cantilever's maximum extension represent how the length between successive ruptures was calculated. A histogram of this length can be seen on the right with a peak position of 9.5 ± 0.5 nm using a Lorentzian fit. (Reproduced with permission from reference 3. Copyright 2003.)

Possible explanations for this short length could be the result of interdomain interactions, domain intermediates, or a specific protein-surface interaction. Interdomain interactions were eliminated as a candidate by repeating the experiment using 1M guanadinium hydrochloride (GdmHCl), a reagent documented to minimize these forces.[12] More extended protein conformations were observed in the surface images, and a most probable length of 14.5nm was obtained. This slightly longer length is consistent with the denaturant extending the protein on the surface, but it is still much shorter than the value obtained for domain unfolding. We could not eliminate the possibility of an intermediate giving rise to the short length since salt deposition prevailed when using 4M GdmHCl. However, as will be discussed later, the length observed in our single molecule measurements most likely result from a specific surface protein interaction and not the 10 nm intermediate predicted from molecular dynamics simulations.[13]

Mechanical Behavior of Fibronectin Aggregates

Because domain unfolding events were not observed in our single molecule experiments, we explored the possibility that the isolated proteins had undergone partial denaturation due to their surface adsorption. Several groups[14] have observed that an increase in protein concentration stabilizes neighboring proteins and prevents their denaturation, perhaps by protecting the domains from interacting with the surface. Figure 8 shows the results from our aggregate

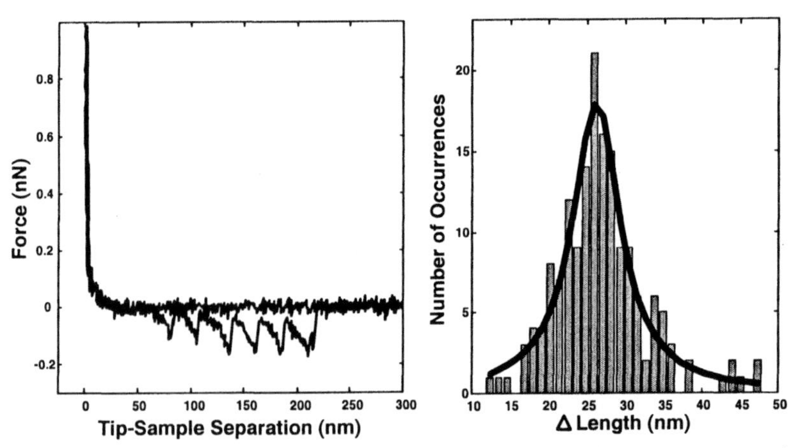

Figure 8: (Left) A force plot obtained for aggregates of FN on mica with the right panel showing a distribution of the lengths between successive ruptures. A peak position of 26.2 ± 0.6 nm, indicative of Type III domain unfolding events was obtained. (Reproduced with permission from reference 3. Copyright 2003.)

studies. We now see a peak shift from 9.5 ± 0.5 nm to 26.2 ± 0.6 nm, a length agreeing with literature reports for Type III domain unfolding events, suggesting aggregation is needed for protein stabilization.

Estimating Protein Stability Using Atomic Force Microscopy

The energy landscape a molecule traverses during its forced extension can be modeled. Several theories have been developed, perhaps the most widely used being the Bell-Evans approach,[15] which allows information about the transition state and the barrier kinetics to be obtained. The basis of the Bell-Evans model involves the rupture force showing a dependence on the rate at which the force was applied to the system, *i.e.* the loading rate (r). As Equation 8 demonstrates, there is a logarithmic dependence of this rupture force (F) on the loading rate.

$$F = \frac{k_B T}{x_\beta} \ln\left(\frac{x_\beta r}{k_{off}^{\circ} k_B T}\right) \tag{8}$$

Here x_β represents the distance from the free energy minimum to the transition state, and k_{off} is the rate of unfolding under zero applied force (F). When collecting data over a range of loading rates, a plot of rupture force *versus* the ln(r) will reveal linear regions, each region corresponding to a barrier overcome by the system during its forced extension. Complications can arise though with Equation 8 when studying flexible polymers and proteins since these systems do not display linear elastic properties. However, other groups[16] have shown a linear dependence of the rupture force *versus* the logarithm of the loading rate when unfolding proteins. Figure 9 shows two of the results obtained in our studies.

Panel A represents single molecule studies, excluding the last rupture in the force plots which again represents protein-tip detachment, and Panel B represents force plots containing *only* a single rupture. The Bell-Evans model gives parameters nearly identical for these two analyses indicating that our single molecule experiments with a length between ruptures of ~9.5 nm involves a specific protein surface interaction and does not represent a domain intermediate being observed. Furthermore, domain unfolding events show only one barrier being overcome in their forced extension and subsequent unfolding (data not shown).

It is also worth noting that the values obtained for the distance from the free energy minimum to the transition state projected along the direction of applied force (x_β) using the Bell-Evans approach are unrealistic for the second linear region in Figure 9. A more realistic barrier position, height, and reaction kinetic information could be obtained using a newer model reported in the literature by Hummer and Szabo (data not shown).[17]

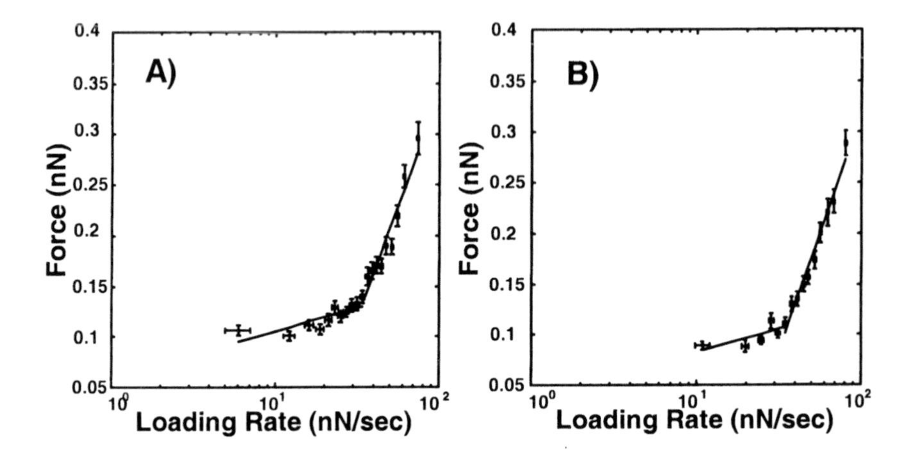

Figure 9: Force vs. loading rate plots for single molecules of FN on mica. Plot A includes all ruptures except the final one while Plot B contains only force plots with a single rupture, indicative of protein-tip detachment. Values obtained from the Bell-Evans model for x_β, and $\Delta G(0)$ are as follows: (Left): 0.20 ± 0.05 nm, 16.8 ± 0.6 kcal/mol; 0.021 ± 2E-3 nm, 14.7 ± 0.28 kcal/mol and (Right): 0.21 ± 0.095 nm, 16.2 ± 1.0 kcal/mol and 0.020 ± 1.2E-3 nm, 14.6 ± 0.18 kcal/mol. Because of the similarities between these plots, it is most likely that the single molecule studies were probing a specific surface-protein interaction. (Reproduced with permission from reference 3. Copyright 2003.)

Therefore, when FN is adsorbed on negatively-charged surfaces, AFM has demonstrated that aggregation is essential for domain unfolding events to be observed. Plots of the length between successive ruptures, rupture force, and loading rate plots confirm that isolated, single molecules on the surface are already partially denatured prior to their extension by AFM.

Acknowledgments

We gratefully acknowledge financial support from ONR (N00014-02-1-0327) and NSF (PHYS-0103048).

References

1. Bemis, J.E.; Akhremitchev, B.B.; Walker, G.C. *Langmuir*, **1999**, 15, 2799.
2. Al-Maawali, S.; Bemis, J.E.; Akhremitchev, B.B.; Leecharoen, R.; Janesko, B.G.; Walker, G.C. *J. Phys. Chem. B*, **2001**, 105, 3965.
3. Meadows, P.Y.; Bemis, J.E.; Walker, G.C. *Langmuir*, **2003**, 19, 9566.
4. Johnson, K.L.; Kendall, K.; Roberts, A.D. *Proc. R. Soc. London, Ser. A*, **1971**, 324, 301.
5. Marko, J.F.; Siggia, E.D. *Macromolecules*, **1995**, 28, 8759. Rief, M.; Gautel, M.; Oesterhelt, F.; Fernandez, J.M.; Gaub, H.E. *Science*, **1997**, 276, 1109. Rief, M.; Oesterhelt, F.; Heymann, B.; Gaub, H.E. *Science*, **1997**, 275, 1295.
6. Yoshizaki, T.; Yamakawa, H. *Macromolecules*, **1980**, 13, 1518.
7. Fleer, G.J.; Stuart, M.A.C.; Scheutjens, J.M.H.M.; Cisgrove, T.; Vincent, B. *Polymers at Interfaces*; Chapman & Hall: London, 1993.
8. Stuart, M.A.C.; Scheutjens, J.M.H.M.; Fleer, G.J. *J. Polym. Sci.*, **1980**, 18, 559. Fleer, G.J.; Scheutjens, J.M.H.M.; Stuart, M.A.C. *Colloids Surf.*, **1988**, 31, 1. Dobias, B. *Coagulation and Flocculation, Theory and Applications*; Marcel Dekker: New York, 1993.
9. Cooper, A.R. *Determination of Molecular Weight*, John Wiley and Sons: New York, 1989; Chapter 10.
10. Oberdorfer, Y.; Fuchs, H.; Janshoff, A. *Langmuir*, **2000**, 16, 9955.
11. Oberhauser, A.F.; Marszalek, P.E.; Erickson, H.P.; Fernandez, J.M. *Nature*, **1998**, 393, 181.
12. Khan, M.Y.; Medow, M.S.; Newman, S.A. *Biochem. J.* **1990**, 270, 33.
13. Gao, M.; Craig, D.; Vogel, V.; Schulten, K. *J. Mol. Bio.*, **2002**, 323, 939.
14. Norde, W.; Favier, J.P. *Colloids Surf.* **1992**, 64, 87. Grinnell, F.; Feld, M.K. *J. Biomed. Mater. Res.*, **1981**, 15, 363. Steadman, B.L.; Thompson, K.C.; Middaugh, C.R.; Matsuno, K.; Vrona, S.; Lawson, E.Q.; Lewis, R.V. *Biotechnol. Bioeng.* **1992**, 40, 8.
15. Bell, G.I. *Science*, **1978**, 200, 618. Evans, E. *Annu. Rev. Biophys. Biomol. Struct.* **2001**, 30, 105.
16. Rief, M.; Gautel, M.; Oesterhelt, F.; Fernandez, J.M.; Gaub, H.E. *Science*, **1997**, 276, 1109. Carrion-Vasquez, M.; Oberhauser, A.F.; Fowler, S.B.; Marszalek, P.E.; Broedel, S.E.B.; Clarke, J.; Fernandez, J.M. *Proc. Natl. Acad. Sci. U.S.A.*, **1999** , 96, 3694.
17. Hummer, G.; Szabo, A. *Biophys. J.*, **2003**, 85, 5.

Chapter 12

Application of Probe Microscopy to Protein Unfolding: Adsorption and Ensemble Analyses

Richard Law, Nishant Bhasin, and Dennis Discher*

Biophyscial and Polymer Engineering Lab, Chemical and Biomoleculer Engineering, University of Pennsylvania, 112 Towne Building, Philadelphia, PA 19104

Single molecule force probe microscopy enables precise mechanical measurements on polymers and proteins that have been first adsorbed to a substrate. Extensibility is indeed central to biological function for many proteins such as the spectrin superfamily of cytoskeletal proteins that consist of 3-helix repeated structures. Here we present recent ensemble-scale analysis of force probe measurements on several spectrin constructs in the context of initial efforts to account for surface adsorption effects. Saturable adsorption of protein is first demonstrated and used to extract a surface interaction energy for spectrin on mica. Surface probing with 1000's of AFM-contact extensions show sawtooth patterns that peak at relatively low forces (< 50 pN) and with one or more (tandem) repeat unfolding events. 1D kinetic modeling of protein unfolding processes in competition with desorption show that surface adsorption effects delimit experimental datasets by introducing extension-terminating processes related to adsorption energies. In sum, protein adsorption may be a prerequisite for extensible unfolding of proteins by AFM, but desorption is readily modeled by simple expressions that fully capture ensemble-scale statistics of these types of single protein unfolding experiments.

Introduction

Atomic force microscopy has been widely used in various single molecule studies including biotin-avidin bond-rupture force measurements (1), the elastic response of double and single-stranded DNA molecules (2), the extensibility of random flight biopolymers (3), and the unfolding of various multi-domain proteins (4-8). In general, the measurements involve extending molecules that are either on or near a surface (e.g. mica), enabling the characterization of stretching and perhaps transition forces such as unfolding. Results reveal an energy landscape that underlies these processes in ways that are not otherwise accessible via other techniques. Protein unfolding, in particular, increasingly explains cell-scale phenomena ranging from stress-facilitated cell adhesion attachment (9) to cytoskeletal viscoelasticity (10,11). Extensible unfolding of single proteins by AFM was initially reported to yield similar kinetic parameters to solution unfolding by denaturants (12), but subsequent mutagenesis studies have suggested different pathways, hence different kinetics (13). Not considered to date, however, has been the relationship(s) of forced unfolding to protein adsorption and substrate interactions.

Spectrin family proteins have multiple tandem repeats of extensible three-helix folds and are thus highly suitable for AFM measurements of extensible unfolding. Forces to unfold are known to be relatively low (3,4) but still compete successfully with forced desorption. Moreover, the spectrin network that crosslinks actin filaments in the red cell is well-known for its resilient elasticity (14), although the sub-molecular basis for network softness is not completely clear. The high homology between α-actinin's four repeat rod domain and at least the four repeat, actin-binding ends of red cell αI- or βI-spectrins strongly suggests contiguous helices between tandem erythroid repeats. Such helical linkers are intriguing in their implications for spectrin extensibility and red cell elasticity, because force-propagated unfolding from one repeat to the next might explain both thermal-softening (15) and strain-softening of the intact erythrocyte network (11,16).

Here we applied the AFM method of single molecule mechanics to purified red cell αI- or βI- spectrin monomers. The goal is to study unfolding pathways in sufficient statistical detail to begin commenting on competition with adsorption/desorption. Although the work focuses on spectrin, the aim is to further develop general characteristics of single molecule AFM measurements in order to facilitate future experimental work with different proteins and different surfaces or AFM tips. Following adsorption studies, full analyses of the thousands of force spectrograms for each protein are done by first categorizing each according to the number of peaks as well as the sawtooth patterns' peak-to-peak lengths and force amplitudes. From scatterplots of force versus length, a common force scale for unfolding both single and multiple (tandem) repeats is clarified together with single and multi-chain unfolding events. Relative frequencies of unfolding states or pathways are also explored in order to begin simple modeling with explicit inclusion of substrate adsorption effects. Experimental results compare well to simulations that are extended to longer spectrin chains as generally found in nature.

Experimental Methodologies

Protein Preparation

The four N-terminal repeats of the 17 domain erythroid βI-spectrin ($\beta_{1\text{-}4}$) and the last four repeats of αI-spectrin ($\alpha_{18\text{-}21}$) (Table 1) were expressed recombinantly and prepared as described (17). Two and three repeat truncations of these two constructs were also studied. Monomer was purified by gel permeation chromatography in phosphate-buffered saline (PBS) and kept on ice for AFM studies. Purified α- or β-spectrin constructs exist only as monomers in solution. Immediately before use, any protein aggregates were removed by centrifugation at 166,000g at 2°C for 1 hour; dynamic light scattering was used to verify monodispersity prior to experiments.

An AFM experiment began by adsorbing 0.03–0.1 mg/ml protein (50 μL drop) for 15 min at room temperature onto either freshly cleaved mica. The surface was then lightly rinsed with PBS and placed without drying, under the head of the AFM; all measurements were carried out in PBS. Lower protein concentrations generated minimal AFM results; higher protein concentrations showed higher unfolding forces, proving consistent with the conclusions below which indicate that domains in multiple, parallel chains will be forced to unfold all at once. Fluorescence imaging of labeled spectrin demonstrated homogeneous adsorption to the surface, and AFM imaging after scratching the surface further showed that no more than a monolayer of molecules covered the substrate (8). Fluorescence adsorption to various surfaces was done before the AFM experiments by using an ammonolysis reaction to label protein with the 5-(and-6)-carboxytetramethylrhodamine, succinimidyl ester. Free dyes were separated from labeled proteins using centrifugation filters and chromatography. The labeled proteins were allowed to adsorb to mica under the same experimental conditions for AFM as described above.

Dynamic Force Spectroscopy

Two AFMs were used with similar results: (i) a Nanoscope IIIa Multimode AFM (Digital Instruments, Santa Barbara, CA) equipped with a liquid cell and (ii) an Epi-Force Probe from Asylum Research. Sharpened silicon nitride (SiN_3) cantilevers (Park Scientific, Sunnyvale, CA) of nominal spring constant k_C = 10 pN/nm were commonly used, with equivalent results obtained using 30 pN/nm cantilevers. k_C was measured for each cantilever by instrument-supplied methods, and additional calibrations were performed as described previously (8). Experiments (at ~23°C) were typically done at imposed displacement rates of 1 nm/msec as well as at 0.5 and 5 nm/msec. For any one speed, thousands (eg. 6000-7000) of surface to tip contacts were generally collected and later analyzed with the aid of a custom, semi-automated, visual analysis program. For a many hour experiment, initial results compared very favorably with results obtained near the end of the experiment.

Results and Discussion

Surface adsorption effects in protein unfolding

Protein adsorption and desorption are key processes in AFM experiments on extensible unfolding of proteins. For the AFM tip to successfully pick up and unfold the protein, the protein must adsorb to both the substrate and the AFM tip. To investigate spectrin adsorption, we performed a fluorescence adsorption study on the four repeat $\beta_{1\text{-}4}$ and the two repeat $\beta_{1\text{-}2}$ constructs. The intensity or adsorbed mass for various concentrations of the labeled proteins was measured using fluorescence microscopy. Figure 1 plots the adsorbed mass/intensity versus concentration and shows that $\beta_{1\text{-}2}$, the smaller sized construct, adsorbs less to the substrate compared with the larger sized $\beta_{1\text{-}4}$. The graphs fit very well to the saturable Langmuir adsorption isotherm. The adsorption constant K for $\beta_{1\text{-}4}$ was found to be smaller (0.09 mg/ml) than that for $\beta_{1\text{-}2}$ (0.5 mg/ml) indicative of stronger interactions in adsorption of $\beta_{1\text{-}4}$ with the substrate.

Protein adsorption factors include molecular properties of the protein, microenvironment (e.g. buffer, salt), substrate properties (charge), and time allowed for adsorption. Since the last three factors were the same for both constructs, the adsorption differences reflect molecular differences in protein size, structure, charge, etc (*18*). The larger sized $\beta_{1\text{-}4}$ construct obviously has larger surface area for contact with more subunits to interact with the mica substrate, providing some rationale for the smaller K (tighter interactions).

Table 1 Percentage of spectrograms with analyzable unfolding curves (i.e $N_{pk} \geq 3$) out of 6000-7000 contacts.

Construct	*Spectograms with $N_{pk} \geq 3$*
α_{20-21}	4.9 %
β_{1-2}	2.8 %
β_{1-3}	3.0 %
β_{1-4}	8.3 %

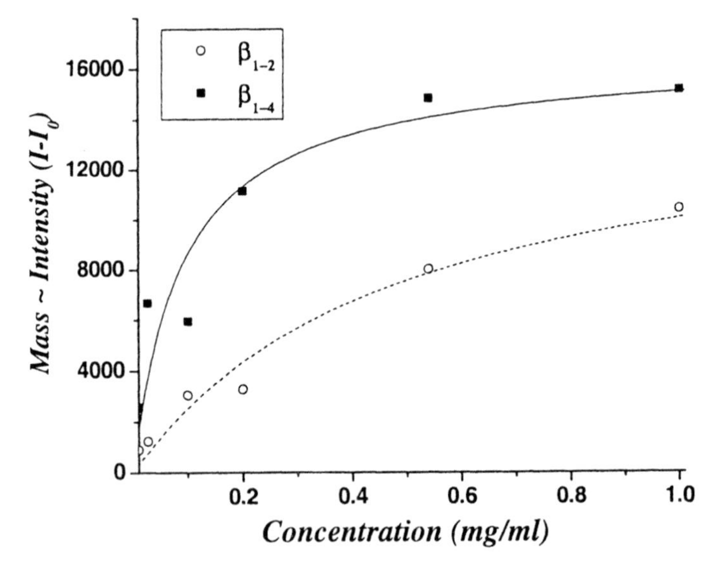

Figure 1. Adsorbed mass, a, versus bulk concentration, c. Both the β_{1-4} and β_{1-2} were pre-labeled with the 5-(and-6)-carboxytetramethylrhodamine-succinimidyl ester, yielding the same ratio of dye:protein mass. The intensity (adsorbed mass) of the labeled protein was measured using fluorescence microscopy. β_{1-2}, the smaller size protein adsorbed less to the mica substrate compared with the larger size β_{1-4}. The experiments were fit by Langmuir adsorption isotherms, $a = Ac/(K+c)$, yielding $K = 0.09$ mg/ml and 0.5 mg/ml for β_{1-4} and β_{1-2} respectively. AFM extension of protein was done near $c \approx K$.

Sawtooth patterns of force extension spectrograms

Here surfaces coated with two, three-, and four-repeat constructs of spectrin monomers (Figure 2A) were probed using sharpened AFM tips, with similar surface concentrations of protein studied by using bulk concentrations of 0.2 mg/ml of $\beta_{1\text{-}2}$ and 0.1 mg/ml of $\beta_{1\text{-}4}$ construct (i.e. ~K or less). Subsequent approach and retract curves under the AFM tip are sometimes very distinctive. The sawtooth patterns from tip retractions shown in Figure 2B are recognized as unique to protein unfolding since other materials such as polymers, DNA, and RNA all reportedly show simpler, monotonic, stretching patterns (19-21). AFM-imposed extension of two, three, and even four domain constructs of spectrin yield similar forced unfolding patterns. The exponentially increasing portions of the asymmetric force–extension curves are fit well by the worm-like chain (WLC) model for entropic elasticity and correspond, as is typical in such experiments, to extension of an unfolded domain up to the point where another (hitherto folded) domain in the chain unfolds.

Figure 2B shows two representative extension curves from room temperature experiments; one with five peaks (N_{pk} = 5) showing nearly constant length intervals and the other with four peaks (N_{pk} = 4) showing one extra long length interval. Despite the difference in the number of peaks, both force 'spectrograms' yield similar total extension length before desorption from tip or substrate. The one extra long length interval appears to be twice the length of the other intervals and indicates tandem unfolding events (i.e. two or more repeats unfolding simultaneously) (4). More extensive analyses of many such curves consistently show the factor of two. The last spectrogram in Figure 2B illustrates how a 4 peak spectrogram is analyzed. The first peak and the height of the last peak are ignored in the experimental analyses as desorption events, and peak forces and peak-to-peak ($l_{pk\text{-}pk}$) lengths were analyzed as illustrated. The total unfolding length is the sum of all the $l_{pk\text{-}pk}$, and this sum is found to never exceed the molecule's total contour length (4).

The finding that two repeat constructs, $\alpha_{20\text{-}21}$ and $\beta_{1\text{-}2}$, can be extensibly unfolded by AFM appears particularly noteworthy. Unfolding of two domains is surprising in that a minimal amount of polypeptide is available for adsorption to both tip and substrate. Regardless of construct length, the force-extension spectrograms here reproduce several key features already found for a range of monomeric spectrin chains (4,20). First, the heights of the force peaks are again found to be less than ~50 pN which is clearly much smaller than the 150 – 300 pN forces originally reported for titin under similar rates of protein extension (0.01–10 nm/msec) (3). The force-scale for spectrin unfolding appears independent of substrate effects since similar forces had already been obtained with spectrin physisorbed to both gold and mica substrates (4); similar results are also found with amino-silanized glass (positively charged) even though

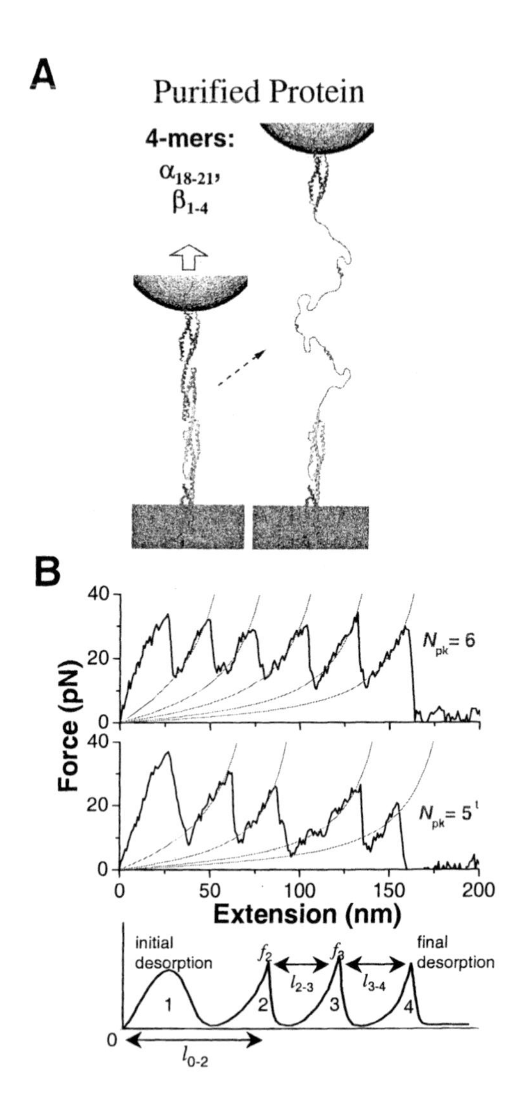

Figure 2. Forced extension and unfolding of single molecules by AFM. (A) A single chain is shown adsorbed to both an AFM tip and a rigid substrate. Folded ribbon structures are those of α-actinin's rod domain (Ylanne et al., 2001). As a folded chain is tensed and a domain unfolds, the applied force relaxes and the chain extends, producing one sawtooth in a pattern of force relaxation. The right shows one such repeat mostly unfolded and sketched as suggested by implicit-solvent molecular dynamics simulations of Paci and Karplus (2000). (B) Sketch of an $N_{pk} = 4$ spectrogram. The first peak and the height of the last peak are ignored in experiment analyses as desorption events. Peak forces and peak-to-peak lengths that are analyzed are indicated. The total unfolding length is the sum of all the indicated $l_{i\text{-}j}$.

protein adsorption is about 5-fold larger than on mica (22). Given the relatively low isoelectric points of these spectrin constructs (i.e. net negative charge) (17), the AFM results appear consistent with forced unfolding rather than strictly forced desorption. One caveat on this is the fact that all experiments to date have used silicon-nitride AFM tips (though with various k_C).

Ensemble analyses of AFM extension

To assess the relative frequencies of unfolding pathways versus competitive stochastics of the desorption process, it is critical that unbiased analyses be applied. This is appropriate because these AFM methods of protein extension are intrinsically random. For a 'successful' contact, an AFM tip comparable in radius to the size of an entire protein must come into contact with unseen molecule(s) pre-adsorbed to the substrate (Figure 2A). The contact occurs, of course, at an unknown position or domain on the molecule. Second, in pulling the AFM tip away from the substrate, the molecule or molecules should bridge the gap between tip and substrate, but a long enough molecule could also form a loop that maintains two points of contact with either substrate or tip. Even at low protein concentrations, looped chains seem unavoidable for sufficiently long molecules. Third, since a molecule is only physisorbed to the AFM tip (and the substrate also in the present studies), continued pulling will stress physisorbed attachment(s) and tend to disrupt them as modeled later. Desorption could certainly occur before all possible domains between tip and substrate have been conformationally distended or unfolded. Marszalek et al. (23) allude to some of these effects in pulling on polysaccharides that, like proteins, exhibit conformational transitions under force. With mixtures of polysaccharides, they point out that: "On the average, 1 out of 10 trial contacts between the AFM tip and the substrate results in a force spectrogram that is unambiguously interpretable. In the other 9 trials we picked up no molecules at all (the most common case) or we picked up too many molecules of similar length that produced a complicated, uninterpretable spectrogram." Polysaccharides tend to give less complicated, monotonically increasing spectrograms than multidomain proteins, so an analysis of more rather than less data can be a challenge. However, we show here that a thorough analysis can also provide important insights into both the existence and frequency of novel, less frequent, states and pathways of protein unfolding as well as desorption limits. Arguably, unless one knows ahead of time what force spectrogram is expected, any other selective sorting of single molecule data can introduce significant bias.

Peak distributions from force-extension curves

To characterize and categorize the data from a given experiment consisting of thousands of force-extension events, a first statistical analysis involved generating an N_{pk} distribution (Figure 3). Distributions of N_{pk} offer insight into the unfolding pathways of the protein under force (i.e. partial or multi-step unfolding) as well as highlight the random nature of AFM experiments, including terminal desorption. Percentage occurrences for each N_{pk} are indicated in each bar of the histograms of Fig. 3. AFM force contacts in PBS alone yield predominantly $N_{pk} = 0$ (99.6 %) with a small few $N_{pk} = 1$ (0.4 %) curves. The presence of protein is thus clearly indicated by any extension curves whenever $N_{pk} \geq 2$. Correlating the N_{pk} distributions with the number of domains D in a given spectrin construct, we have found that for a given construct with D domains, the maximum number of peaks, $max(N_{pk})$, increases linearly with D as:

$$max(N_{pk}) = D + 2 \quad (1)$$

Each repeat thus gives exactly one or fewer peaks. Proteins that unfold in multi-steps (partial) would have a stronger increase with D than in Equation 1. The y-intercept of Equation 1 shows that curves with $N_{pk} = 2$ correspond to protein detachment from the surface (first peak) and then subsequent detachment from either the tip or the surface (last peak). The first peak is usually the most symmetric and the single peak in buffer alone (Fig. 3A) further supports the idea that the first peak represents initial desorption of the chain or tip rather than unfolding. Based on these general interpretations of Eq. 1, only $N_{pk} \geq 3$ are analyzable unfolding events, and these are given as cumulated percentages in Table 1. At the low protein concentrations used (~ K or less), these events are seen to be a minor percentage (<10%) of total contacts. In the context of molecular interactions, it has been established that experiments are predominantly in the single molecule limit when the frequency of events is 25% or less (24). The percentages in Table II are well within this limit and thus appear consistent with the analyzable sawtooths being predominantly (though not exclusively) single molecule experiments.

Unfolding lengths and force distributions

The peak-to-peak unfolding lengths, $l_{pk\text{-}pk}$, and unfolding force peak distributions (Figure 4) provide the most succinct summary of the thousands of force-extension sawtooth patterns that constitute an experiment. The peak-to-peak unfolding length histograms are bimodal with principal means differing by a factor of (exactly) 2.0. The major peaks in the length distribution were fitted with sums of Gaussians that reflect proportional contour lengths for single repeats (Table 2) through sums of multiple Gaussians using l_c-proportioned

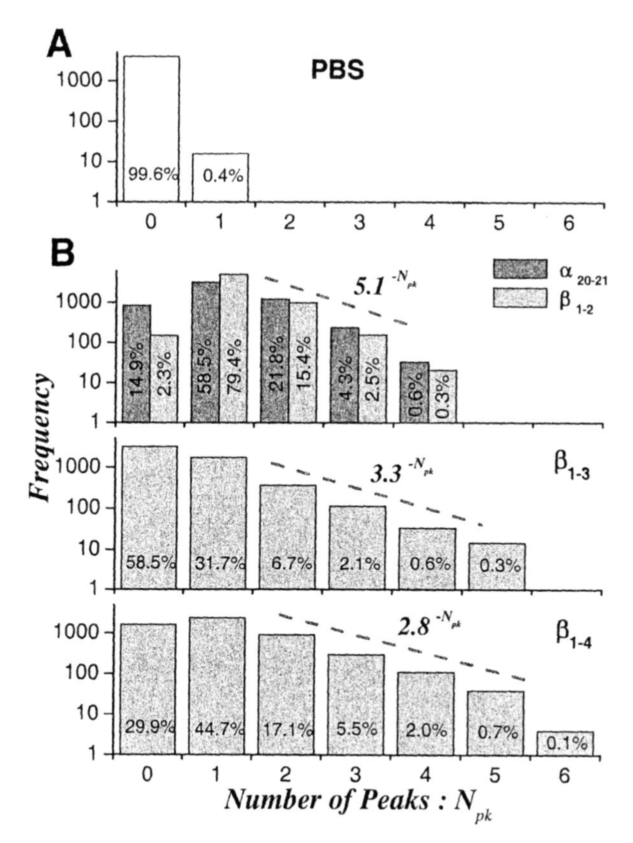

Figure 3. Number of peaks, N_{pk}, per extension in (A) PBS buffer alone without any protein, or (B) with pre-adsorbed spectrin constructs containing 2, 3, or 4 repeats. Each experiment involved 5000 contacts between AFM tip and substrate. For PBS alone, more than one peak was never seen. For spectrin, exponential decays of the form m^{-Npk} generally fit the multi-peak results between $1 < N_{pk} < max(N_{pk})$ with the indicated values for m. This figure is courtesy of Law et al. (4).

Table 2. Primary structure properties of spectrin constructs (aa = amino acid). Domain and total contour lengths, l_O and Lc respectively, have been calculated with a peptide length of 0.37 nm

Domain	*Number of aa*	*Contour, l_O (nm)*	*Domain*	*Number of aa*	*Contour, l_O (nm)*
α21	111	41.0	β1	120	44.3
α20	114	42.1	β2	114	42.1
α19	107	39.5	β3	109	40.2
α18	109	40.2	β4	106	39.1
extra	9	3.3	extra	2	0.7
Total, L_C		166.1			166.4
AVERAGE		41.5			41.6

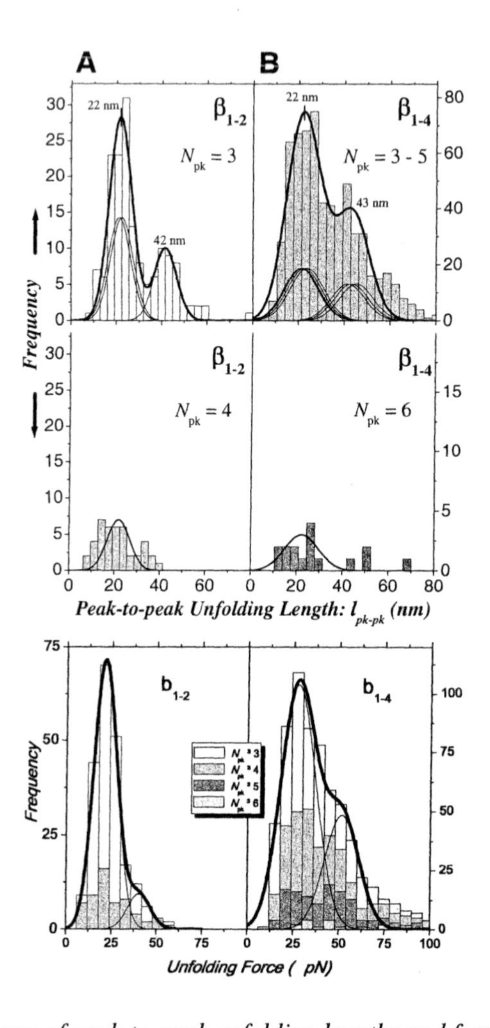

Figure 4. Histograms of peak-to-peak unfolding lengths and forces for sawtooth extensions of (A) $\beta_{1\text{-}2}$-spectrin and (B) $\beta_{1\text{-}4}$-spectrin. The upper two bimodal histograms show results for $N_{pk} < max(N_{pk})$. The major peaks were fitted with sums of Gaussians that reflect proportional contour lengths for single repeats (see Table 2). The minor peaks were likewise fitted but with contour lengths of tandem repeats. The overall sum of all the Gaussians is indicated by the heavy black line; a factor of about two between the major and minor peaks is apparent. The lower two histograms show results only for $N_{pk} = max(N_{pk})$ and are found to be mostly enveloped by the major peaks, i.e. single repeat peaks, of the upper histograms. The lowest two histograms show the unfolding forces for sawtooth extensions. The bimodal histograms were fitted with two Gaussians of the same width, and the overall sums of the Gaussians are indicated by heavy black lines. This figure is courtesy of Law et al. (4).

means (8). With $D = 2$ (Figure 4A) for example, single repeat unfolding was accounted for in $N_{pk} = 3$ spectrograms by summing two equal-height Gaussians with means constrained by the ratio l_{c1} / l_{c2}; tandem repeat unfolding was accounted for by simultaneous fitting of a third Gaussian. For $D = 2$ and $N_{pk} = 3$, the bimodal fit yields a major peak at 22 nm and a more minor peak at 42 nm. The minor peaks were likewise fitted but with contour lengths of tandem repeats. The major peaks in the unfolding distributions correspond to unfolding a single repeat, the freely-fitted factor of 2.0 in mean length between the first and second peaks provides a strong indication of unfolding a tandem pair of in-series repeats. In addition, the $N_{pk} = 6$ monomer histogram shows that the majority (90%) of the $l_{pk\text{-}pk}$ fall within the major envelope of single repeat unfolding events, consistent with $N_{pk} = 6$ corresponding to four single repeat events plus initial and final desorption events. The few events outside of the single repeat envelope evidently reflect over-extension of the cumulated slack that develops toward the end of a long extension before final desorption terminates the spectrogram (25).

Since the lengths of the three helices in repeats are known from crystal structures to be 4-5 nm or 12-15 nm per repeat, average unfolding lengths of 22–24 nm are only about $^2/_3$ of the adjusted mean contour lengths; 41 nm contour lengths (Table I) and thus indicate an unfolded domain is not greatly stretched from a random coil or molten globule state before the unfolding of the next domain occurs. Implicit-solvent molecular dynamics simulations of Paci and Karplus (26) also suggest helices may persist in unfolding as sketched in Fig. 2A (27). Such a result is consistent with both broader distributions and lower forces for unfolding spectrin in comparison, for example, to titin whose unfolding lengths approach theoretical contour lengths at 5–10-fold higher forces. For spectrin, however, the freely-fitted factor of nearly 2.0 between the major and minor length peaks also provides a strong indication of $^2/_3$–unfolding for a tandem pair of in-series repeats.

Force distributions, cumulated from the heights of the force peak patterns of Figure 2B are also bimodal with Gaussian means differing again by about a factor of ~2. The bimodal histograms were fitted with two Gaussians of the same width. The major Gaussian fit most likely reflects the force to unfold a single repeat or a tandem repeat within a chain, since the force required for tandem unfolding appear to be the same as that required to unfold a single repeat (as explained further below). The minor Gaussian fit at twice the force reflects the unfolding of two chains. To verify this, scatterplots of Figure 5A, pair a given unfolding length with the appropriate unfolding force. These plots clearly suggest that the factor of 2 in force reflects unfolding in either two chains, or a loop in a chain rather than unfolding tandem repeats versus single repeats.

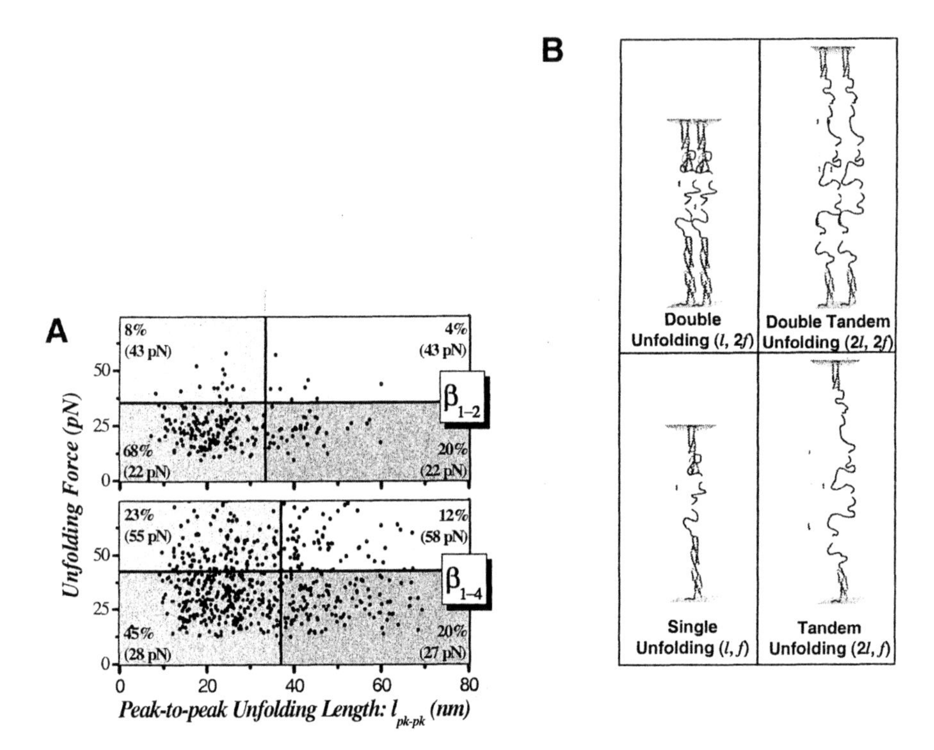

Figure 5. Protein unfolding scenarios from the scatterplot analysis. (A) Scatterplots of unfolding force versus unfolding length. Dividing lines are obtained from the intersection between the major and minor Gaussians of force and length distributions and thus represent a stringent if simplistic separation of unfolding processes or states. The fraction of data points within each of the four quadrants is indicated as a percentage of total data points together with the quadrant-average force in parentheses. This figure is courtesy of Law et al. (4). The four unfolding scenarios represented in the scatterplots are illustrated in (B). Interpretation of these states were made in terms of unit unfolding length, l, and unit unfolding force, f, of a single repeat. The lower two quadrants correspond to single chain unfolding, whereas the upper two quadrants correspond to double chain unfolding of both single and tandem repeat unfolding.

Scatterplots of force-length pathways

By pairing a given unfolding length with the appropriate unfolding force from force-extension curves in Figure 2B, scatterplots shown in Figure 5A prove most revealing of the multiple unfolding processes that underlie the experiments. The scatterplots are divided into four quadrants or states from the bimodal Gaussian intercepts of Fig. 4 and each state represents a different unfolding scenario. All of the four quadrants contain a significant number of unevenly distributed data points. The relative probability, average forces, and lengths were calculated for each quadrant. As with the bimodal distributions for length (Figure 4), the two left-most quadrants differ in average length from the two right-most by a factor of two. This ratio again corresponds to the unfolding of single versus tandem repeats.

Forces in the upper two quadrants of the scatterplots are also seen to be essentially 2.0-fold higher than the forces in the lower two quadrants. The force scale of the upper two quadrants further corroborates the unfolding on either two separate chains or a loop of a single chain that spans the gap between tip and surface as hypothesized in the previous discuss based on the bimodal force distribution. Again, since previous solution studies indicate monomeric chains and a lack of self-association (17), loops are also unlikely to be self-interacting, i.e. repeats in one 'leg' of the loop should respond independently of repeats in the other 'leg'. The unfolding scenarios for each of the four quadrants are illustrated in Fig. 5B.

Comparing the scatterplots of the four repeat β_{1-4} and the two repeat β_{1-2} constructs, differences in the overall data quantity and the percentages of data points in each quadrant are clear. The four-repeat β_{1-4} shows many more data points and more frequent two-chain unfolding (35% versus 12%) compared to the two-repeat β_{1-2} construct. These disparities could be attributed to adsorption effects described earlier (Figure 1) and/or subtle structural differences that affect the unfolding patterns for both constructs. Since spectrin is known to have a persistence length of at least several nanometers (e.g., Lee and Discher (21)) and each triple helical repeat is likely to be stiff, the two domain construct of length ~10-15 nm is undoubtedly stiffer and less likely (with bending resistance) to form a loop compared to the four domain construct. Conversely, longer chains are expected to give more loop results.

The relative probabilities of states indicated in Fig. 5 are also revealing. As expected from the low protein concentrations used in these studies, the number of times that two chains (or a loop of one chain) are pulled upon is small in comparison to a single chain. The lower frequency of tandem repeat unfolding, in either one chain or two, is also consistent with rudimentary statistical expectations. Given four domains to unfold, no more than two pairs can unfold per chain, and so a relative occurrence of about 2:1 single:tandem repeat unfolding is a reasonable first guess. The scatterplots suggest that this simple estimate applies. Nonetheless, in defining a free energy difference by taking the log_n of probability ratios and multiplying by k_BT, all such differences are calculated to be less than ~1 k_BT. Since accessible transition states define the critical points of choice for a system to follow one pathway or another, the $<k_BT$ difference implies that both single and tandem repeat unfolding have very similar transition state energies.

Simulations of unfolding with desorption

To simulate 'domain-by-domain' unfolding of individual domains in a specific multi-domain protein, each domain needs to resist unfolding up to a characteristic force. Sustained force accelerates the process by lowering the energy barrier between the native and unfolded state and also tends to make the stretched state more stable. This unfolding process in spectrin can be further transmitted through successive domains due to the cooperative behavior of the helically linked tandem structure. In addition to such intra-protein interactions, the entire process occurs because the protein adsorbs to the surface (and tip) by attractive interactions that can also clearly be disrupted by force (Figure 2B).

In order to model the mechanical stability of (1) the protein domains and (2) the surface attractions, both are assumed to have a free energy profile described as a function of a single spatial coordinate in extension. One-dimensional energy landscapes of folding/unfolding are then used in kinetic Monte Carlo methods already developed for forced unfolding (3,4). The probability of unfolding for each of the domains is calculated at each time step in a simulation using $P_{unfolding} = \Delta t / t^*$, with Δt specified (to keep $P_{unfolding} < 1$) and t^* dependent on force per the "Bell" factor $exp(-f\, x_u / k_b T)$ (28). x_u is a predetermined length scale for spectrin unfolding (=1.2 nm) (4). Desorption of the chain was similarly modeled with a separate length scale, x_{des}. The argument $(-f\, x)$ has the general role of decreasing the energy barrier under force. Each domain starts in a native state and transforms to an unfolded state(s) depending on the transition probabilities calculated at every subsequent interval for extension. Tandem unfolding is also sampled stochastically. A sample simulation starts with picking up several domains (d_i, $i = 1...D$) and a random number generator picks one of the d_i domains for a transition while competing with chain desorption.

Parameters x_{des} (= 0.9 nm) and a common force-free rate for either unfolding or desorption α_o (= 4×10^{-6} msec^{-1}) were adjusted until the simulated sawtooth patterns yielded frequency ratios that match the respective experiments for $\beta_{1\text{-}4}$ spectrin (Fig 6A). Figure 6B shows the simulated frequency plots of $N_{pk} \geq 2$ for various spectrin constructs analogous to Figure 3B. The N_{pk} distributions consistently show a decay with N_{pk} that are again fit to $\sim m^{-N_{pk}}$. The factor m

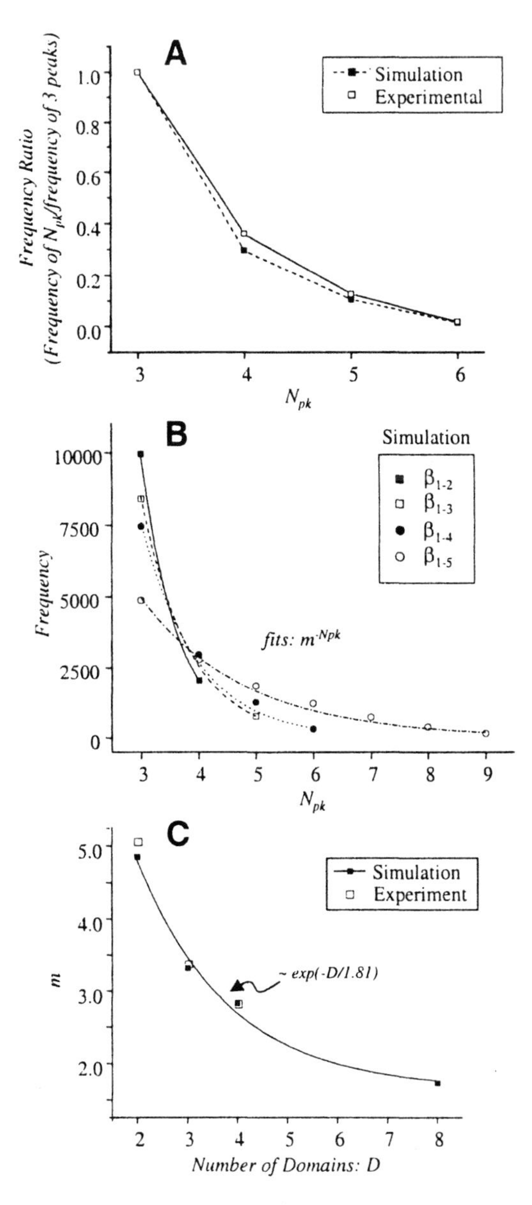

Figure 6. N_{pk} frequency per extension from Kinetic Monte Carlo Simulation versus experiment. (A) Results for the frequency ratio (occurrences with $N_{pk} \geq 3$)/(occurrences with 3 peaks plotted for $\beta_{1\text{-}4}$ protein). (B) N_{pk} frequency plots for each protein show decays of the form m^{-Npk} analogous to Fig. 3B. (C) Plot of m versus number of domains fit with a single exponential.

provides a measure of the m-fold fewer ways of achieving one more unfolded domain (single and tandem) spanning the gap between tip and surface.

The factor m undoubtedly reflects a random desorption process, but it also clearly shows a trend with the total number of domains D (Figure 6C): m decreases exponentially from ~5 to ~2 as D increases from 2 to 8. This decrease in m with D provides a simple measure of the increased number of ways of unfolding domains when more domains, or degrees of freedom, are present. A simple calculation also suggests this: with $D = 2$, there are at most three unfolding pathways since unfolding can involve one tandem repeat event or, in lieu of this, two single repeat events. With $D = 4$, in contrast, there are not only four possible single repeat events but also a combination of these with three tandem repeat events.

Finally, one can estimate a characteristic desorption force based on x_{des} and the Langmuir adsorption isotherm fits of Figure 1. Fitted K (in M) for both β_{1-2} and β_{1-4} in Figure 1 are used to calculate free energies for protein adsorption via $\Delta G = -kT \log K$. Respective calculations for β_{1-2} and β_{1-4} yield characteristic forces for desorption of $f_{des} \sim |\Delta G| / x_{des} = 21$ pN and 25 pN. These forces are comparable to chain unfolding forces, and thus indicate why desorption competes stochastically with unfolding. Additionally, a slightly higher force of desorption for β_{1-4} versus β_{1-2} is consistent with a very reasonable frequency for full unfolding of β_{1-4}.

Summary

As shown here, single molecule force probe microscopy enables precise mechanical measurements on a protein such as spectrin with as few as two domains or repeats (~230 amino acids). The protein must first be adsorbed to a substrate prior to contacting with the AFM tip, and while reproducible and data-rich experiments - rather than statistically starved experiments - are done at surface concentrations well below saturation of fluorescently labeled protein, saturable coverage can be clearly demonstrated. This is ultimately used to extract a surface interaction energy for spectrin on mica that can be used to understand desorption forces. Once the protein is physisorbed, AFM probing with thousands of contact extensions provides an impartially analysed set of sawtooth patterns in which the first peak (probably) and the last peak (certainly) are desorption peaks. Force peaks between the first and last for spectrin occur at relatively low forces of <50 pN at the typical extension rates of 1 nm/msec used in these types of experiments. The length between these peaks is bimodal with a factor of exactly two between means, which indicates that tandem repeats unfold in addition to single repeats. Despite the multiple pathways of forcibly unfolding these spectrin constructs, each repeat clearly contributes a maximum of one peak. This is deduced from ensemble-scale distributions of the number of peaks seen in the thousands of sawtooth patterns collected for each construct. These experiments are nonetheless "designed to fail": successful unfolding of repeats is rare (<10% for three or more peaks indicative of unfolding) and certainly well within upper limits for single molecule success rates. However, scatterplots of force and length demonstrate sufficient sampling because of a significant but separable fraction of two chain results. Furthermore, through 1-D kinetic Monte Carlo modeling of protein unfolding in competition with desorption processes, surface adsorption effects are shown to delimit the datasets in a predictable way. In sum, protein adsorption is certainly a pre-requisite for extensible unfolding of proteins by AFM, but desorption is readily modeled, we have show here, by simple expressions that fully capture ensemble-scale statistics of these types of forced unfolding experiments on single proteins.

Future experiments might focus on the structural state of the protein after it has adsorbed to the substrate. Circular dichroism methods, for example, might be useful for such assessments, but the sawtoothed spectrograms here and elsewhere already provide a relatively clear indication of a folded state. More challenging and interesting is structurally assessing the process of contacting the protein with an AFM tip and extending it while the protein is still in contact at the ends because this almost certainly alters the folded state of the protein at the contact points. Assessing the structure of these adsorbed portions appears far more challenging to assess. Simple models of desorption, as here, offer at least some insight into the energetics involved.

Acknowledgements

We thank George Liao for invaluable assistance with the experiments as well as data analysis. This work was supported by NIH, NSF, and MDA grants.

References

1. Lo, Y.; Simmons, J.; Beebe, T.P. Temperature dependence of the biotin-avidin bond-rupture force studied by atomic force microscopy. *Journal of Physical Chemistry* **2002,** 106, 9847-9852.
2. Smith, S.; Cui, Y.; Bustamante, C. Overstretching B-DNA: The elastic response of individual double-stranded and single-stranded DNA molecules. *Science* **1996**, 271, 795-799.
3. Rief, M.; Gautel, F.; Oesterhelt, F.; Fernandez, J.M.; and Gaub, H.E. Reversible unfolding of individual titin immunoglobulin domains by AFM. *Science* **1997**, 275, 1295-1297.
4. Law, R.; Carl, P.; Harper, S.; Dalhaimer, P.; Speicher, D.W.; Discher, D.E. *Biophys. J.* **2003,** 84, 533-544.

5. Furuike, S.; Ito, T.; Yamazaki M. Mechanical unfolding of single filamin A (ABP-280) molecules detected by atomic force microscopy. *FEBS Lett.* **2001**, 498(1), 72-5.
6. Oberhauser, A.F.; Marszalek, P.E.; Erickson, H.P.; Fernandez J.M.The molecular elasticity of the extracellular matrix protein tenascin. *Nature* **1998**, 393(6681), 181-185.
7. Oberdorfer, Y.; Fuchs, H.; Janshoff, A. Conformational analysis of native fibronectin by means of force spectroscopy. *Langmuir* **2000**, 16, 9955-9958.
8. Carl, P., Kwok; C.H., Manderson; G.; Speicher; D.W.; Discher, D.E. Forced unfolding modulated by disulfide bonds in the Ig domains of a cell adhesion molecule. *Proc. Natl. Acad. Sci. USA*; **2001**, 98, 565-1570.
9. Krammer A.; Craig, D.; Thomas, W.E., Schulten, K.; Vogel, V. A structural model for force regulated integrin binding to fibronectin's RGD-synergy site. *Matrix Biol.* **2002,** 21(2),139-47.
10. Minajeva, A.; Kulke. M.; Fernandez, J.M.; & Linke, W.A. Unfolding of titin domains explains the viscoelastic behavior of skeletal myofibrils. *Biophys J.* **2001**, 80(3), 1442-51.
11. Lee, J.C-M.; Discher, D.E. Deformation-enhanced fluctuation in the red cell skeleton with theoretical relations to elasticity, connectivity, and spectrin unfolding. *Biophys. J.* **2001**, 81, 3178-3192.
12. Carrion-Vazquez, M.; Oberhauser, A.F.; Fowler, S.B.; Marszalek, P.E.; Broedel, S.E.; Clarke, J.; Fernandez, J.M. 1999; Mechanical and chemical unfolding of a single protein: A comparison *Proc. Natl. Acad. Sci. USA*, **2001**, 96 (7), 3694-3699.
13. Fowler, S.B.; Best, R.B.; Toca Herrera, J.L.; Rutherford, T.J.; Steward, A.; Paci, E.; Karplus, M.; Clarke, J; Mechanical unfolding of a titin Ig domain: structure of unfolding intermediate revealed by combining AFM, molecular dynamics simulations, NMR and protein engineering. *J. Mol. Biol.* **2002**, 322, 841-849.
14. Mohandas, N.; Evans, E.A. Mechanical properties of the red cell membrane in relation to molecular structure and genetic defects. *Annu. Rev. Biophys. Biomol. Struct.* **1994**, 23, 787-818.
15. Waugh, R., and Evans, E.A. Thermoelasticity of red blood cell membrane. *Biophys. J.* **1979**, 26, 115-131.
16. Markle, D.R.; Evans, E.A.; and R.M. Hochmuth. Force relaxation and permanent deformation of erythrocyte membrane. *Biophys. J.* **1983**, 42, 91-98.
17. Ursitti, J.A.; Kotula, L.; DeSilva, T.M.; Curtis, P.J., Speicher, D.W. Mapping the human erythrocyte β-spectrin dimer initiation site using recombinant peptides and correlation of its phasing with the α-actinin dimer site. *J. Biol. Chem.* **1996**, 271, 6636-6644.
18. Proteins at interfaces. ACS. **1987**.

19. Yang, G.; Cecconi, C.; Baase, W.; Vetter, I.; Breyer, W.; Haack, J.; Matthews, B.; Dahlquist, F.; Bustamante, C; Solid-state synthesis and mechanical unfolding of polymers of T4 lysozyme; *Proc. Natl. Acad. Sci. USA* **2000**, 97, 139-144.
20. Lenne, P.-F.; Raae, A.J.; Altmann, S.M.; Saraste, M.; and Horber, J.K.H. States and transitions during forced unfolding of a single spectrin repeat. *FEBS Lett.* **2000,** 476, 124-128.
21. Rouzina, I.; Bloomfield, V.A. Force-Induced Melting of the DNA Double Helix. 1. Thermodynamic Analysis. *Biophysical Journal.* **2001**, 80, 882-893.
22. Kwok, C. H. Nanomechanical characterization of skeletal proteins by atomic force microscopy (AFM). M.S.E. Master Thesis, Chemical Engineering, University of Pennsylvania, 2000.
23. Marszalek, P.E.; Li, H.; and Fernandez, J.M. Fingerprinting polysaccharides with single-molecule atomic force microscopy. *Nature Biotechnology* **2001,** 19, 258-262.
24. Shao J-Y.; Hochmuth, R.M. Mechanical Anchoring Strength of L-Selectin, 2 Integrins, and CD45 to Neutrophil Cytoskeleton and Membrane. *Biophys. J.* **1999**, 77, 587-596.
25. Liu, S., Palek; J., Prchal, J.; Castleberry, R. *J. Clin. Invest.* **1981**, 68, 597-605.
26. Paci, E.; Karplus, M. Unfolding proteins by external forces and temperature: The importance of topology and energetics. *Proc. Natl. Acad. Sci. USA* **2000**, 97, 6521-6526.
27. Yip, C.; Yip, C.; Ward, M. *Biochemistry* **1998**, 37, 5439-49.
28. Bell, G.I. Models for the specific adhesion of cells to cells. Science1978, 200, 618-627.

Chapter 13

The Use of Atomic Force Microscopy in Characterizing Ligand–Receptor ($\alpha_5\beta_1$ Integrin) Interactions

Efrosini Kokkoli and Anastasia Mardilovich

[1]Department of Chemical Engineering and Materials Science, University of Minnesota, Minneapolis, MN 55455

A biomimetic system was used to study interactions of the $\alpha_5\beta_1$ receptor with its ligand with an atomic force microscope (AFM). Bioartificial membranes, which display peptides that mimic the cell adhesion domain of the extracellular matrix protein fibronectin, are constructed from peptide-amphiphiles. A novel peptide-amphiphile was designed that contains both GRGDSP (Gly-Arg-Gly-Asp-Ser-Pro, the primary recognition site for $\alpha_5\beta_1$) and PHSRN (Pro-His-Ser-Arg-Asn, the synergy binding site for $\alpha_5\beta_1$) sequences in a single peptide formulation, separated by a spacer. The strength of the PHSRN synergistic effect depends on the accessibility of this sequence to $\alpha_5\beta_1$ integrins. The interaction measured with the immobilized $\alpha_5\beta_1$ integrins and GRGDSP peptide-amphiphiles is found to be specifically related to the integrin-peptide binding. It is affected by divalent cations in a way that accurately mimics the adhesion function of the $\alpha_5\beta_1$ receptor. The dissociation of single $\alpha_5\beta_1$-GRGDSP pairs under loading rates of 1-305 nN/s revealed the presence of two activation energy barriers in the unbinding process. The high-strength regime above 59 nN/s maps the inner barrier along the direction of the force. Below 59 nN/s a low strength regime appears with an outer barrier and a much slower transition rate that defines the dissociation rate (off-rate) in the absence of force (k_{off}^{o}=0.015 s^{-1}).

Introduction

The ability of cells to bind various membrane and extracellular matrix (ECM) proteins through integrins expressed on cell membranes provides signals that affect the morphology, motility and survival of cells (*1,2*). Better understanding of the molecular mechanisms of integrin-ligand interaction could therefore illuminate several important cell functions. Integrins are membrane glycoproteins with two noncovalently associated subunits, designated α and β. The combination of α and β subunits determines the specificity for extracellular ligands as well as intracellular signaling events (*3,4*). Although it is unknown how these subunits associate, it is thought that they exist in different conformations according to their activation status: either inactive, where they are unable to bind ligand, or active, where they are capable of binding their ligand (*3*). In addition, a cell responds to its environment in part through its ability to regulate the function of integrins by modulating their activity. Both divalent cations and monoclonal antibodies have been used as stimuli to investigate the dynamic regulation of integrin-ligand binding (*3*).

In general, ligand binding occurs through recognition by the integrin of a short amino acid sequence from the ligand (*5*). The prototype for these integrin binding sites is the Arg-Gly-Asp (RGD) sequence that is present in fibronectin, fibrinogen, vitronectin, and other adhesive proteins (*5,6*). Short peptides containing the RGD sequence can mimic the cell adhesion domains of proteins in two ways: when they are coated on a surface they promote cell adhesion, whereas in solution they can saturate the capacity of the receptor to bind cell adhesion ligands.

This study presents a way to design a biologically active membrane-like surface in which ligand accessibility and template composition is used as a means to control the interaction with immobilized $\alpha_5\beta_1$ integrins. Bioartificial membranes, supported on solid substrates, displaying peptides that mimic the cell binding domain of the ECM protein, fibronectin, are constructed from mixtures of peptide-amphiphiles and polyethylene glycol (PEG) amphiphilic molecules. Peptide-amphiphiles feature C_{16} dialkyl ester tails, a Glu (glutamic acid) linker, a $-(CH_2)_2-$ spacer and a headgroup that incorporates the bioactive sequence (*7*). The tail serves to align the peptide strands and provide a hydrophobic surface for self-association and interaction with other hydrophobic surfaces.

The two sequences used in this study are found in the ECM protein, fibronectin: the tenth type III module, GRGDSP (the primary recognition site for $\alpha_5\beta_1$), and the ninth type III module, PHSRN (the synergy binding site for $\alpha_5\beta_1$). These two sequences are separated by 30-40 Å in fibronectin (*8*). We have designed a new sequence – PHSRN$(SG)_3$SGRGDSP (referred as PHSRN-GRGDSP in the remaining of the text) – that contains both the primary and the synergy adhesion-mediating sequences of fibronectin in a single peptide formulation. A linker $(SG)_3S$ is used to link these two bioactive sequences in a specific spatial orientation. The theoretical length of this linker is calculated using two different approaches and was found to be approximately 30 Å; 26 Å when calculated using 3.7 Å per amino acid residue (*9*), and 32 Å when calculated using standard bond lengths and angles. Serine (S) and glycine (G) are used here because they are small, uncharged amino acid residues with serine being hydrophilic and glycine hydrophobic. Thus, the goal in designing the linker was to mimic as closely as possible the distance and the hydrophilicity between the PHSRN and GRGDSP sequences in the fibronectin molecule.

A mixture of the peptide amphiphiles and PEG lipid molecules is deposited on a surface by the Langmuir-Blodgett technique. An AFM is used to provide high resolution images and direct adhesion measurements. We have also monitored the single-molecule interaction of $\alpha_5\beta_1$-GRGDSP pairs under a range of loading rates to investigate the presence of multiple transitions in the unbinding pathway. The spontaneous dissociation (off-rate) reaction rate was also calculated for the $\alpha_5\beta_1$ from the force spectroscopy experiment, as this is a parameter of prime interest for any biological ligand-receptor system, and results are in agreement with off-rates reported for $\alpha_5\beta_1$ integrins.

Our results demonstrate that our biomimetic system can give an insight into the biophysical character of unbinding processes between $\alpha_5\beta_1$ receptor-GRGDSP ligand pairs and allow us to understand how different environmental conditions and multiple peptides can enhance the performance of functionalized interfaces.

Materials and Methods

Preparation and Characterization of Bioartificial Membranes

1,2-Distearoyl-sn-glycero-3-phosphatidylethanolamine (DSPE) and the PEG chains with molecular weight of 120 covalently linked to DSPE (PEG120) were obtained from Avanti Polar Lipids, Inc. (Alabaster, AL). $(C_{16})_2$-Glu-C_2-KAbuGRGDSPAbuK, $(C_{16})_2$-Glu-C_2-GRGESP, $(C_{16})_2$-Glu-C_2-PHSRN, and $(C_{16})_2$-Glu-C_2-PHSRN$(SG)_3$SGRGDSP, were synthesized as described elsewhere (*7*). For simplicity these peptide-amphiphiles will be referred as GRGDSP, GRGESP, PHSRN, and PHSRN-GRGDSP respectively, in the rest of the text.

The pure amphiphiles were dissolved at approximately 1mg/ml in a 99:1 chloroform/methanol solution. The solution was stored at 4°C and heated to room temperature prior to use. We used the Langmuir-Blodgett (LB) technique to create supported bioactive bilayer membranes. LB film depositions were done on a KSV 5000 LB system (KSV Instruments, Helsinki, Finland). All the depositions were done at a surface pressure of 41 mN/m, which is well below the collapse pressure of 60 mN/m. The deposition speed for both the up and down strokes was 1 mm/min. The first step in producing a supported bilayer membrane was to make the mica hydrophobic with a layer of DSPE in the upstroke. The second layer with the peptide-amphiphiles was deposited in the down stroke. The resulting supported bilayer membranes were transferred into glass vials under water. Care was taken to avoid exposing the surfaces to air, as they rearrange to form monolayers and trilayers (*10*).

AFM characterization of the LB films was done with a Digital Instruments Nanoscope III system equipped with a fluid cell for tapping mode (Digital Instruments, Santa Barbara, CA). Images were obtained in tapping mode under water using standard 100 μm V-shaped silicon nitride AFM cantilevers with pyramidal tips (Digital Instruments).

$\alpha_5\beta_1$ Immobilization

Purified human $\alpha_5\beta_1$ integrins were purchased from Chemicon International (Temecula, CA). The integrins were dialyzed overnight at 4°C against solution

A, pH 7.2, containing 20 mM Tris·HCl (Sigma), 0.1% Triton X-100 (Sigma), 1mM $MgCl_2$ (Aldrich), 1mM $MnCl_2$ (Aldrich) and 0.5 mM $CaCl_2$ (Aldrich). The integrins were removed from the dialysis membrane, diluted to 5 μg/ml in solution A, aliquoted and stored at -80°C. The integrin immobilization was performed on the AFM cantilevers as described elsewhere (*11,12*).

AFM

Surface force measurements were performed using a commercial AFM, a Nanoscope III (Digital Instruments, Santa Barbara, CA), in contact mode in 1 mM $MnCl_2$ solution (unless stated otherwise), at a loading rate, defined as the spring constant of the cantilever times the velocity of the piezo, of 59.8 nN/s, using standard 200 μm V-shaped silicon nitride AFM cantilevers with pyramidal tips (Digital Instruments) of nominal radius 5-40 nm and nominal spring constant 0.06 or 0.32 N/m. Data were recorded as the two surfaces, the sample surface and the probe tip, were brought into contact and then pulled apart. The adhesion force or the "pull-off" force is defined as the minimum force required to separate two surfaces. All adhesion force measurements were carried out at room temperature and at pH of $MnCl_2$ solution of 6.4 – 6.5. In order to minimize the drift effects, the AFM was warmed up for at least half an hour before an experiment. AFM force data were converted to force-distance curves using the DI-AFM software (Nanoscope III).

For the single-molecule experiments three different kinds of V-shaped silicon nitride AFM cantilevers were used with nominal spring constants of 0.01, 0.03 and 0.06 N/m (Digital Instruments, Santa Barbara, CA and Thermo-microscopes, Sunnyvale, CA). The loading rate was varied, by changing both the spring constant of the cantilever and the velocity of the piezo, from 1-305 nN/s.

Results and Discussion

Figure 1 shows a schematic representation of the experimental system used in this study.

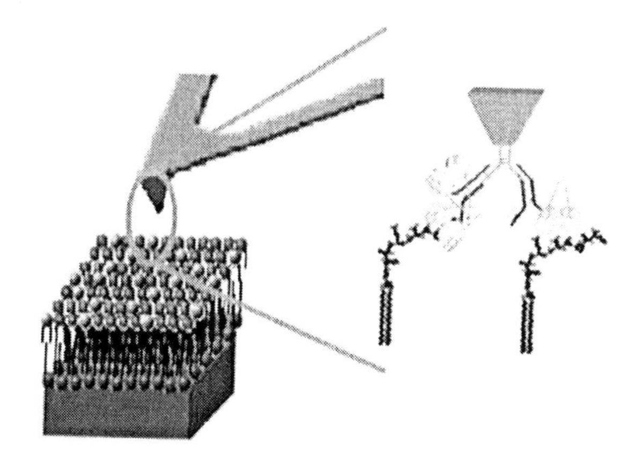

Figure 1. Schematic representation of the experimental system, not drawn to scale. Two different antibodies were used to immobilize and activate purified human $\alpha_5\beta_1$ integrins on the AFM tips. Bioartificial membranes were deposited on mica surfaces. Forces were measured in a liquid environment with the AFM.

Specificity was confirmed in two ways: by varying the ion concentration (Figure 2) and by using inactive sequences, such as GRGESP peptide-amphiphile and PEG120 (Figure 3), as negative controls. The role of divalent cations in integrin function is demonstrated by the lack of ligand binding upon removal of cations by chelating agents (*13*). Furthermore, divalent cations such as Mn^{2+}, Mg^{2+} and Ca^{2+} have distinct effects on integrin function in vitro. Generally, Mn^{2+} confers high affinity binding properties on isolated integrins, whereas Ca^{2+} inhibits ligand binding, with Mg^{2+} playing a stimulatory role but to a lesser extent than Mn^{2+} (*3*). The effect of Mn^{2+} and Ca^{2+} ions on the immobilized $\alpha_5\beta_1$ integrins is shown in Figure 2. The same tip was used for all these measurements. Mn^{2+} ions increase the adhesion of the immobilized $\alpha_5\beta_1$ to the GRGDSP ligands, whereas Ca^{2+} ions decrease it. Decrease of integrin activity is shown by addition of EDTA, that chelates cations. To show that this effect is reversible at the end of the experiment the AFM cell was flushed with 5 mM $MnCl_2$ and the adhesion that was measured was the same magnitude as the one present at the beginning of the experiment for the same concentration.

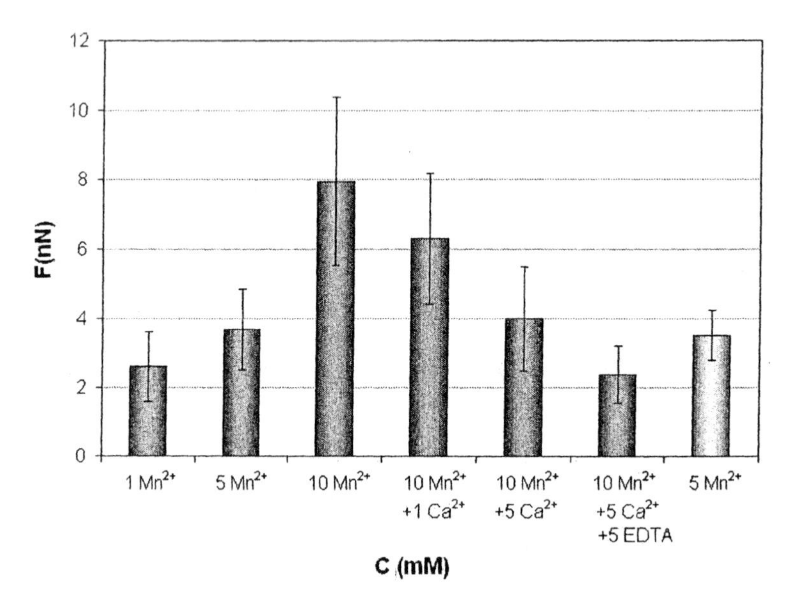

Figure 2. Effect of divalent cations on the specific binding of $\alpha_5\beta_1$ – GRGDSP (12). The same tip was used for these measurements. Each column is the average of 25-35 measurements. The error bars show standard deviations and reflect the fact that the number of pairs that interact every time the two surfaces are brought into contact can vary from one measurement to another. (Reproduced from reference 12. Copyright 2004 American Chemical Society.)

The RGD motif in fibronectin is the critical recognition site for $\alpha_5\beta_1$, but the synergy site, PHSRN, is also required for high affinity binding (*14*). This high affinity binding was tested with bioartificial membranes that had 50%GRGDSP 50%PHSRN peptide-amphiphiles and surfaces that had mixtures of the new sequence PHSRN-GRGDSP and PEG120. AFM images of these surfaces verified that they are well mixed. The PEG120 has a much shorter headgroup compared to the PHSRN-GRGDSP headgroup and thus was used as an effective way to provide more space for the PHSRN-GRGDSP peptide-amphiphile to bend and expose more of the active sequence at the interface (*12*). Our results in Figure 3 demonstrate that the specific recognition of the immobilized receptor was significantly increased for a surface that presented both the primary recognition site (GRGDSP) and the synergy site (PHSRN). $\alpha_5\beta_1$ integrins do not bind to PEG120 and GRGESP amphiphiles which also confirms specificity. Figure 3 shows measurements from two different tips and as can be seen experimental measurements are reproducible.

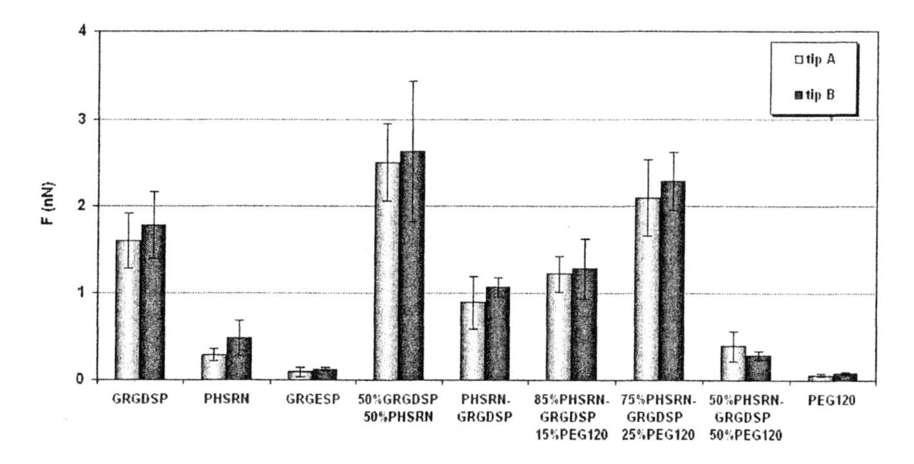

Figure 3. Adhesion measurements between immobilized $\alpha_5\beta_1$ integrins and different surfaces in 1 mM $MnCl_2$ solution. Each column is an average of 30-40 measurements on each surface. The error bars show standard deviations and reflect the fact that the number of pairs that interact may vary from one measurement to another.

Finally the effect of the loading rate on the single-molecule interaction of $\alpha_5\beta_1$ integrins was investigated. For this, force histograms were collected at loading rates of 1-305 nN/s (*11*).

The possibility of membrane failure via extraction of peptide-amphiphiles was investigated by comparing our data to the forces measured for the extraction of lipids from lipid bilayers (*15*). For comparable rates of 1-10 nN/s the force required to extract a lipid is 30-44 pN (*15*) whereas the unbinding force of a single $\alpha_5\beta_1$–GRGDSP pair is 16-19 pN. Therefore, under these conditions lipid anchoring is stronger and the integrin bond will unbind first.

The unbinding force versus the loading rate for single $\alpha_5\beta_1$-GRGDSP bonds is plotted in Figure 4. The force spectrum revealed two regimes with different

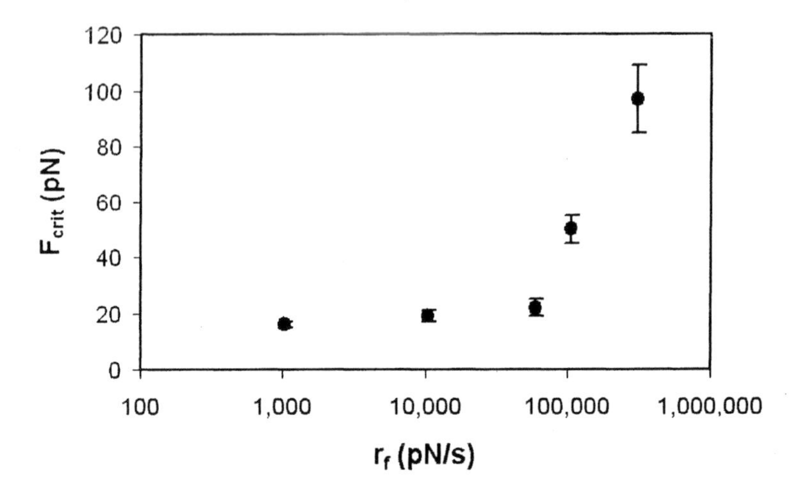

Figure 4. Rupture force, F_{crit}, for $\alpha_5\beta_1$-GRGDSP bond versus the loading rate, r_f, (11). The error bars represent the standard deviations in each force histogram. A total of 4000 individual forces have been analyzed for this graph. (Reproduced from reference 11. Copyright 2004 American Chemical Society.)

slopes, within the range of loading rates that was examined. Thus, we were able to determine that the $\alpha_5\beta_1$-GRGDSP complex overcomes two transitions, an inner and outer barrier, during its dissociation (*11*). The spontaneous dissociation (off-rate) reaction rate, k_{off}^o, was also calculated for the $\alpha_5\beta_1$ from the force spectroscopy experiment, as this is a parameter of prime interest for any biological ligand-receptor system, and was found to be k_{off}^o=0.015 s^{-1} (*11*). A dissociation rate constant of 0.01 s^{-1} has been reported between fibronectin and the fibronectin receptor ($\alpha_5\beta_1$) on fibroblast cells in solution (*16*). A k_{off}^o=0.012 s^{-1} was recently shown from single-molecule AFM measurements between K562 cells, that express the $\alpha_5\beta_1$ integrins, and human plasma fibronectin (*17*). Thus, the k_{off}^o=0.015 s^{-1} reported here from a biomimetic cell-free system that consists of pure immobilized $\alpha_5\beta_1$ receptors and GRGDSP peptide ligands is in excellent agreement with the recently reported value of 0.012 s^{-1} measured from AFM cell-protein interactions and the value of 0.01 s^{-1} from solution cell-protein measurements.

Conclusions

We have engineered a novel biomimetic system that allows us to study the mechanistic details of the unbinding processes of $\alpha_5\beta_1$-GRGDSP pairs and use this complex to understand how different conditions and multiple peptides can enhance the adhesion of $\alpha_5\beta_1$. For example, stronger adhesion can be achieved if the surface is functionalized with peptides that mimic both the main recognition site (GRGDSP) for integrins and the synergy site (PHSRN) of the cell adhesion domain of fibronectin.

The specificity of the $\alpha_5\beta_1$ integrin for the GRGDSP ligand was established by varying the concentration of divalent cations, by using chelating agents and an inactive peptide-amphiphile (GRGESP) or PEG120 as a negative control experiment. It was shown that Mn^{2+} and Ca^{2+} had a distinct effect on the immobilized integrins. As in the case of integrins that are part of a cell membrane, Mn^{2+} increased the adhesion of the immobilized $\alpha_5\beta_1$ for the GRGDSP peptide-amphiphile, Ca^{2+} decreased it and loss of integrin activity was shown by adding the chelating agent EDTA.

The strength of the $\alpha_5\beta_1$-GRGDSP single bond varied between 15 and 109 pN for loading rates 1-305 nN/s. The adhesion measured after surface separation cannot be attributed to membrane failure as the force required to pull lipids from the bilayer is significantly higher than the forces measured with this system for comparable loading rates. Under these conditions lipid anchoring is stronger and the integrin bond will unbind first. Dynamic force spectroscopy revealed that the $\alpha_5\beta_1$-GRGDSP dissociation is characterized by two barriers. The outer barrier is governing the rate of spontaneous dissociation in solution while the inner one becomes rate-limiting in the presence of much higher loading rates. The off-rate was calculated to be 0.015 s^{-1}, which is in agreement with solution and AFM cell measurements for $\alpha_5\beta_1$ integrins. Thus our experiments show that our biomimetic system can accurately predict the solution off-rate parameter, one of the main determinants of the affinity of the integrin-ligand system.

Acknowledgements

Acknowledgment is made to the Donors of the American Chemical Society Petroleum Research Fund, for partial support of this research.

References

1. Hynes, R. O. *Cell* **1992,** *69*, 11-25.
2. Ruoslahti, E.; Reed, J. C. *Cell* **1994**, *77*, 477-478.
3. Fernandez, C.; Clark, K.; Burrows, L.; Schofield, N. R.; Humphries, M. J. *Frontiers BioSci.* **1998**, *3*, 684-700.
4. Giancotti, F. G.; Ruoslahti, E. *Science* **1999**, *185*, 1028-1032.
5. Ruoslahti, E. *Annu. Rev. Cell Dev. Biol.* **1996**, *12*, 697-715.
6. Pierschbacher, M. D.; Ruoslahti, E. *Nature* **1984**, *309*, 30-33.

7. Berndt, P.; Fields, G. B.; Tirrell, M. *J. Am. Chem. Soc.* **1995**, *117,* 9515-9522.
8. Leahy, D. J.; Aukhil, I.; Erickson H. P. *Cell* **1996**, *84*, 155-164.
9. Idiris, A.; Alam, M.T.; Ikai, A. *Protein Eng*. **2000**, *13*, 763-770.
10. Hansma, H. G.; Clegg, D. O.; Kokkoli, E.; Oroudjev, E.; Tirrell, M. *Methods Cell Biol.* **2002**, *69*, 163-193.
11. Kokkoli, E.; Ochsenhirt, S. E.; Tirrell, M. *Langmuir* **2004**, *20,* 2397-2404.
12. Mardilovich, A.; Kokkoli, E. *Biomacromolecules* **2004**, *5,* 950-957.
13. Mould, A. P.; Akiama, S. K.; Hamphries, M. J. *J. Biol. Chem.* **1995**, *270*, 26270-26277.
14. Aota, S.; Nomizu, M.; Yamada, K. M. *J. Biol. Chem.* **1994**, *269*, 24756-24761.
15. Evans, E.; Ludwig, F. *J. Phys.:Condens. Matter* **2000**, *12*, A315-A320.
16. Lauffenburger, D. A.; Linderman, J. J. *Receptors: Models for Binding, Trafficking, and Signaling;* Oxford University Press: New York, NY, 1993; pp 30.
17. Li, F.; Redick, S. D.; Erickson, H. P.; Moy, V. T. *Biophys. J.* **2003**, *84*, 1252-1262.

Polymer Surface Characterization by Scanned Probes

Chapter 14

Following Processes in Synthetic Polymers with Scanning Probe Microscopy

Jamie K. Hobbs[1,*], Andrew D. L. Humphris[2], and Mervyn J. Miles[2]

[1]Department of Chemistry, University of Sheffield, Sheffield S3 7HF, United Kingdom
[2]H.H. Wills Physics Department, University of Bristol, Tyndall Avenue, Bristol BS8 1TL, United Kingdom

The use of a temperature controlled stage, combined with electronic control of the effective quality factor of the atomic force microscope cantilever to allow faster scanning, has enabled real-time measurements of spherulitic growth at the lamellar scale over the full range of crystallisable temperatures in polyhydroxybutyrate-co-valerate (PHB/V). A possible future route to considerably increased scan-rates has also been explored, and imaging rates approaching the millisecond level have been obtained.

One of the key advantages of scanning probe microscopes (SPM) compared to other real-space techniques with similar resolution is the ability to follow processes in real-time in a wide range of different environments. For polymeric materials this ability to follow the kinetics of a process, such as a phase transition, is particularly important, as the long relaxation times of long chain molecules frequently leads to multiple metastable states, so equilibrium considerations are often unhelpful.

Over recent years the applications of SPM and in particular atomic force microscopy (AFM) have increased significantly with the development of high stability temperature stages that allow many more processes, of both scientific and commercial interest, to be accessed. Particular attention has focussed on polymer crystallization *(1-4)* and melting *(5)*, although recently other transitions such as phase separation and de-wetting *(6)* have been studied.

In all cases, it is clearly necessary to obtain multiple images of the same area during the time period over which the process occurs, that is, the time to collect an image has to be significantly shorter than any time-scales of interest within the process under investigation. AFM is a relatively slow technique – data points are collected in series to build up an image, and each individual image typically takes a minute or more to collect – so in many cases the critical timescales are unobtainable. There are two principle factors controlling the speed at which images can be collected. Firstly, the cantilever on which the AFM probe is mounted has to respond to changes in interaction. When working in intermittent contact, this timescale (the relaxation time, τ) is controlled by the quality factor of the cantilever (its 'Q') such that $\tau=2Q/\omega_0$ where ω_0 is the resonant frequency - a low Q cantilever will respond more rapidly than a high Q cantilever with the same resonant frequency. It has been shown by Sulchek *et al (7)* that using an electronic feed back loop to reduce the effective Q of a cantilever can allow somewhat faster image-rates to be obtained. In this article we will primarily discuss the combination of such an electronic Q-damping system with high stability temperature control to allow the lamellar scale crystallization of a spherulite forming polymer to be followed over the full range of crystallization temperatures from the glass up to close to the melting temperature. The second factor limiting imaging speed in conventional AFM is the mechanical nature of the scan stage. We have recently developed a novel scanning probe microscope capable of scanning 1000 times faster than conventional methods *(8)*. A brief outline of this new technique will also be given.

Temperature Controlled AFM

Temperature control stages for atomic force microscopes must fulfil several requirements. Firstly, the piezoelectric scan-stage must not be allowed to reach extreme temperatures, as there is a risk of either de-poling or, through thermal shock, of damaging the ceramic. Secondly, if intermittent contact mode is to be used, a drive mechanism that is stable over a wide temperature range must be used. Thirdly, the sample itself must be held stably, which means that the control over temperature must be sufficient to prevent drift that will degrade the image. Finally, temperature gradients across the sample need to be kept to a minimum.

We have developed a simple system for use with a Veeco Dimension microscope that fulfils these requirements, and has been described in some detail elsewhere *(9)*. A schematic is shown in figure 1. There are two advances in this system with respect to those that we have used previously. Firstly, a peltier device placed in contact with the underside of the Linkam optical microscope heater allows temperatures down to 0°C to be obtained (lower temperatures cause the formation of condensation on the glass windows). In this set-up, the peltier is used in a manner more akin to a cold-finger, being run at constant current and providing a low temperature against which the conventional heater can control at sub-ambient temperatures simply through heating. This method was found to give very stable temperature control.

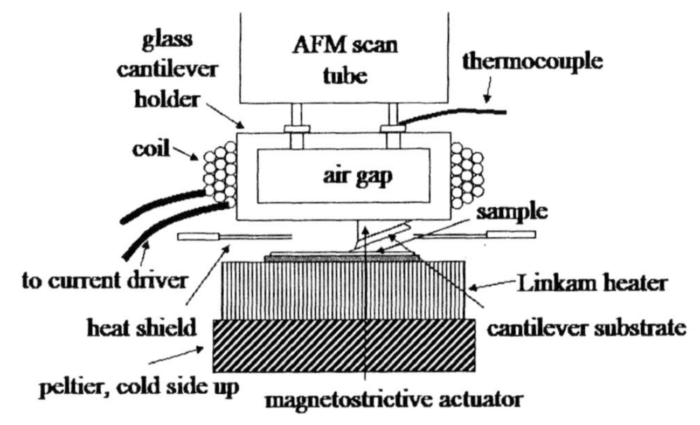

Figure 1. A schematic showing the cantilever and heater system used for temperature controlled AFM.

The second new feature is that a magnetostrictive actuator is used to drive the cantilever, rather than coating the cantilever in a ferromagnetic material. The magnetostrictive material used, Terfenol D, is capable of strains similar to those obtained with piezoelectric materials such as PZT. The reason for this modification is that, by removing the need for magnetic coating, the instrument becomes significantly easier to use, while maintaining the advantage of the magnetic drive that there is no need for electrically (and therefore thermally) conducting links to be made to anything in direct contact with the cantilever. By using a voltage to current converter, the standard drive signal given by the Veeco controller is still used, so there is no other degradation in the usability of the system when compared to the standard microscope.

We have found that using the above set-up allows us to access temperatures in the range 0-190°C with no noticeable change in image quality. As an example of such high stability imaging at elevated temperature, during a temperature ramp, figure 2 shows a series of images taken with this system of polyethylene lamellae being heated up to, and through, the melting point. The sample is being heated at 0.5°C/min. This causes a drift in position of the sample relative to the

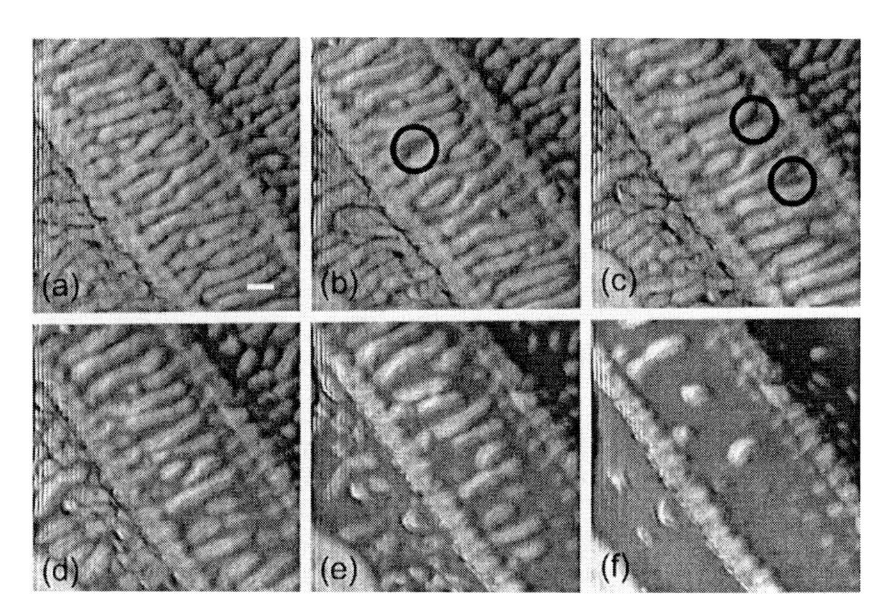

Figure 2. A series of phase images showing the gradual melting of an oriented polyethylene structure, taken during heating at 0.5°C/min. (a) at 126.8°C, (b) at 130.9°C, (c) at 132.8°C, (d) at 134.3°C, (e) at 135.7°C, (f) at 137.6°C. The scale bar represents 100nm.

scanner in a direction from bottom left to top right, at a rate of 2.7nms^{-1}, resulting in a slight distortion in shape of the objects imaged. All the images were collected scanning from bottom to top and at the same heating rate, giving a constant distortion in shape, allowing comparisons between the images to be made with confidence.

It is worthwhile making a few comments on the new insights that can be gained from these images into the melting and high temperature re-organisation of the polymer. The sample was made by dragging a hot razor blade across the molten polymer at a temperature of 140°C and then quenching to room temperature. This preparation method gives highly oriented 'shish-kebab' structures with a largely extended chain backbone, formed during the initial extension of the sample with the razor blade, and lamellar overgrowths what grow during cooling. Quenching the sample means that these overgrowths will have grown over a range of temperatures, with the ends of the lamellae having grown at the lowest temperature and therefore being least stable. From the figure, we can see several instances where short, infilling lamellae melt out (circled in figure 2b and 2c). This melting out of the shorter lamellae seems to be accompanied by a gradual thickening of the remaining lamellar population. However, the lamellar thicknesses measured (65nm in figure 2a, compared to

78nm in figure 2c, from the 2D FFT of the whole image) are rather thicker than expected, most probably due to a combination of tip broadening (this should not, in fact, influence the peak-to-peak distances measured by this method) and surface effects during crystallization reducing the density of lamellae that grow.

Further heating of the samples leads to more melting back from the ends of the lamellae, and apparently further thickening of the remaining lamellar population. Measuring thicknesses of these now isolated lamellae is more difficult however, as tip convolution becomes more important, as well as the possibility of the tip shape changing over time through interaction with the adhesive polymer melt. In the last two images shown there is more widespread melting of the remaining lamellae, surprisingly leaving small blocks of material isolated in between the two oriented backbones. This observation is in contrast to some of our other observations of melting of shish kebab crystals, and is most probably due to the very high densitiy of lamellae initially along the backbone, as evidenced by its nodular texture. Finally the oriented backbones are left virtually isolated in the melt. Other data (not shown) has shown that these oriented structures are stable up until around 145°C.

From the data set shown in figure 2 it is clear that high quality images can be obtained during continuous heating of the sample. However, in order to obtain data such as this, the rate of scanning must be pushed to its limit, and even then the heating rate is rather limited. In the images shown there are vertical lines on the left hand side of the images, caused by the scan-tube starting to resonate as it is shocked during the rapid change of direction at the end of each scan line. This is one of the limitations on scan speed, and can be slightly alleviated by introducing a "rounding" to the scan line, which provides a more gradual change in velocity at the end of each scan line. As discussed above, the second main limitation on scan speed is the response of the cantilever. In the next section the use of electronic control of the resonance characteristics of the cantilever is discussed briefly, and is then applied to the study of polymer crystallization.

Active control of the cantilever resonance for faster scanning

When imaging, the microscope feedback loop adjusts the location of the cantilever in the z axis so as to maintain an approximately constant amplitude. As the cantilever is scanned over the surface, the interaction it feels changes due to changes in topography and material properties. However, as it is necessary to work in dynamic mode when imaging these soft materials, there is energy stored in the oscillations of the cantilever, and it takes some time for it to react to a change in its environment. This time, which will depend on the relaxation time (or settle time) of the cantilever, is dependent on the mechanical properties of the cantilever and on its environment. The relaxation time of a cantilever is proportional to its Quality factor, so in order to scan fast we require a cantilever with a low Q. However, force sensitivity increases with increasing Q, so there is a compromise to be made between high scan speeds and reduced sample damage. We have found that it is useful to electronically tune the resonance of the cantilever, using a feedback technique described in detail elsewhere *(10,11)*. However, in that earlier work, the main objective was to increase the Q for use in a liquid environment. Here we have followed the example in *(7,12)* and reduced the quality factor of the cantilevers so as to reduce its 'settle time' and hence allow faster scanning.

The ActivResonance Controller from Infinitesima Ltd was used to control the effective quality factor of the cantilever. As this is still a relatively new technique, a brief outline of the method will be given. The response signal of the cantilever, as monitored by the split-photodiode, is phase shifted in the ActivResonance Controller by -90° so as to be in phase with the velocity dependent component of the cantilevers response (i.e. the component that is altered by changing the damping applied to the cantilever). This phase-shifted signal is then amplified and fed back into the drive signal for the cantilever. So, the new cantilever drive signal now contains a component that depends on the response of the cantilever, and which is the mathematical equivalent of increasing the effective damping of the cantilever, i.e. of reducing its Q-factor.

Typically the Q of the silicon cantilevers used (nominal spring constant $50Nm^{-1}$, resonant frequency ~250kHz) is of the order 300. We found that reducing this to approximately 100 resulted in the optimum effective cantilever properties for both sensitivity and scan-rate. This value was only optimised for the silicon cantilevers used in this study, and it is envisaged that if cantilevers with a considerably higher spring constant were used, it would be necessary to maintain a higher Q for the sake of force sensitivity. However, this optimum value does not depend on the natural Q of the cantilever, and is rather used as an additional control parameter during imaging which can be optimised for the particular combination of surface roughness and sample 'softness' under study. The sample we have used is a biodegradable thermoplastic, polyhydroxybutyrate-co-valerate (PHB/V), a statistical copolymer with 24% valerate. The relatively high copolymer content considerably reduces the crystallization rate, giving a peak growth rate of only around $50nms^{-1}$, as shown in figure 3. However, the temperature dependence of growth rate follows the familiar bell-shaped curve, and dense, often banded, spherulites are formed over the whole crystallization range studied. This makes it an ideal material for a study such as this. In the following, all images are collected at 256×256 pixels.

Figure 4 shows a pair of images of the growth front of a spherulite of PHB/V, one taken with the ActivResonance Controller feedback electronics switched off, and the other with them switched on and the effective quality factor

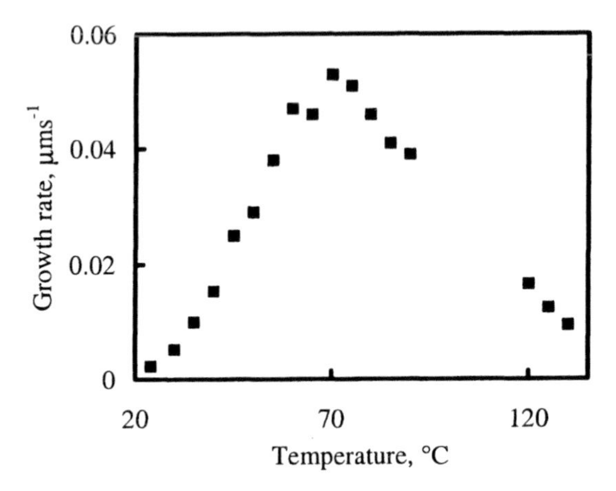

Figure 3. The variation in growth rate with temperature for polyhydroxybutyrate-co-valerate.

reduced by a factor of three. The sample was prepared as described in *(13)* and was imaged under ambient conditions.

These two images were both collected at the limit at which the scan-stage can be used without excessive degradation of the image due to ringing – as it is, the left hand side of each image suffers considerably from the direction change of the scanner. The image obtained with an electronically reduced Q shows significantly better definition, particularly at the end of the lamellae, where the crystalline regions meets the melt. For the purpose of following the kinetics of the process, this is the most important part of the images, so any improvement has value.

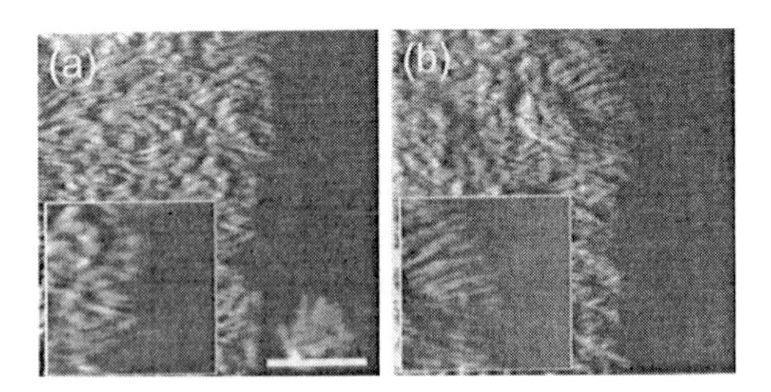

Figure 4. Two AFM phase images of a PHB/V spherulite growth front. (a) taken without electronic control of the effective Q, Q~300, and (b) taken with an effective Q~100 several minutes later on the same sample. The inserts show 0.5µm zooms of the lamellar growth front, from which the improved definition in (b) is clearly visible. Both images collected at a line rate of 27.47 Hz. The scale bar represents 500nm.

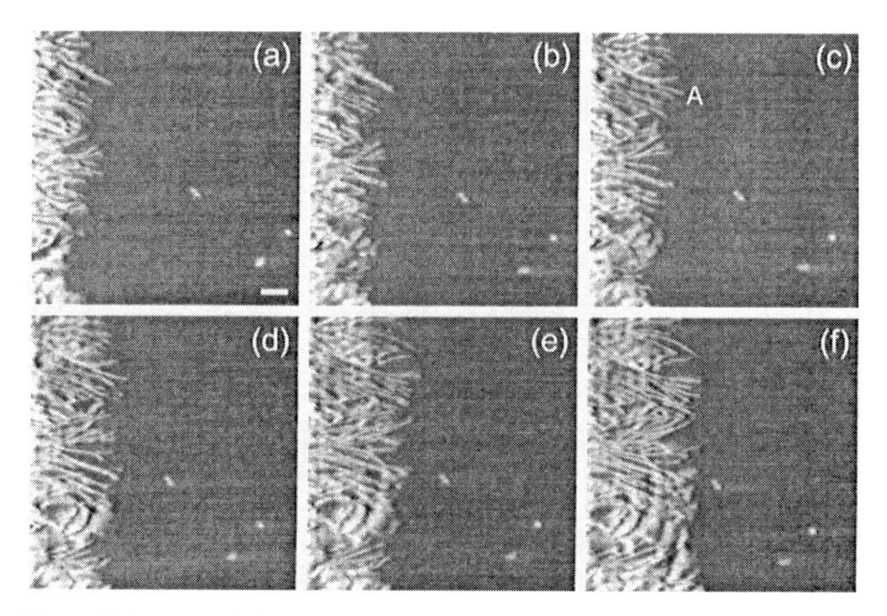

Figure 5. A series of phase images of the lamellar growth front of a spherulite of PHB/V crystallizing at room temperature. The images were collected consecutively at 11.7 second intervals. The effective Q of the cantilever was set to 90, the natural Q was 245. The scale bar represents 100nm.

Figure 5 shows a series of images taken consecutively during crystallization under ambient conditions, each image collected in 11.7 seconds. The fine lamellar structure of the growth front is clearly visible.

Figure 6a shows a cross-section of part of the growth front, showing that the lateral resolution is still at the pixel level. The width of the lamellae of PHB/V is ~5nm, which is in good agreement with the measurements taken from these images. From the images shown here, the growth rates of individual lamellae can be measured over several frames. In a previous study of this same material *(13)* we showed that the growth rate of individual lamellae was not a constant, as predicted by most polymer crystallization theories, but rather varied with time. However, in that work, we were only able to measure the rate of growth of a single lamellar over two or at most three images. Figure 6b shows the growth rate of the lamella marked A in the figure 5, in which the measurements were taken over the series of six images shown. As in that previous work, the growth rate varies over time. Measurements of several lamellae show variations in growth rate from 2 to 13nms^{-1}, which is in good agreement with the less direct measurements made in our previous study of this same material *(13)*. Great care was taken to ensure that these measurements were not influenced from any possible artefacts due to the serial nature of the data collection or any possible differences between consecutive scans by comparison with static objects. However, the fact that the rate of lamellar growth varies from crystal to crystal is irrefutable by simple inspection of adjacent lamellae over several images.

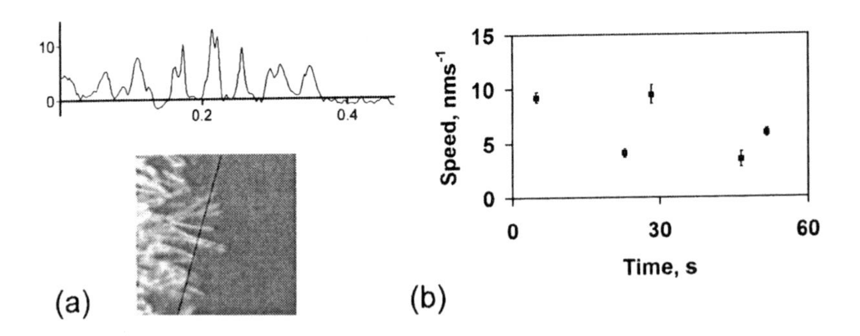

Figure 6. 6(a) shows a cross-section taken from a software zoom of part of figure 5b, the line along which the section is taken being indicated by a line in the small inset figure. The widths of the sharp peaks that represent single lamellae is ~6nm (measured at half height). 6(b) shows the rate of growth of a single lamella taken from the images in figure 5.

Figure 7 shows a series of images of PHB/V growth taken at different temperatures from close to the glass transition temperature up to close to the melting temperature. The increased imaging speeds enabled by use of damping of the effective Q of the cantilever allow sufficient time resolution to make a study such as this useful.

Despite the high scan rates, the image quality is still sufficient to clearly discern the individual lamellae. One of the possible by-products of reducing the effective Q is to reduce force sensitivity and, potentially, increase the damage to the sample surface. However, it is clear from close examination of the growth structure of areas that have been imaged during growth, and areas that have not, that the low effective Q has not led to an increase in damage to the sample. We have not carried out a quantitative analysis of the change in force sensitivity with changing effective Q, and would expect there to be some reduction. The key point here is that, for the current application, the faster scan-rates enabled by use of a reduced effective Q have allowed new insights into the crystallization behaviour of the polymer to be gained.

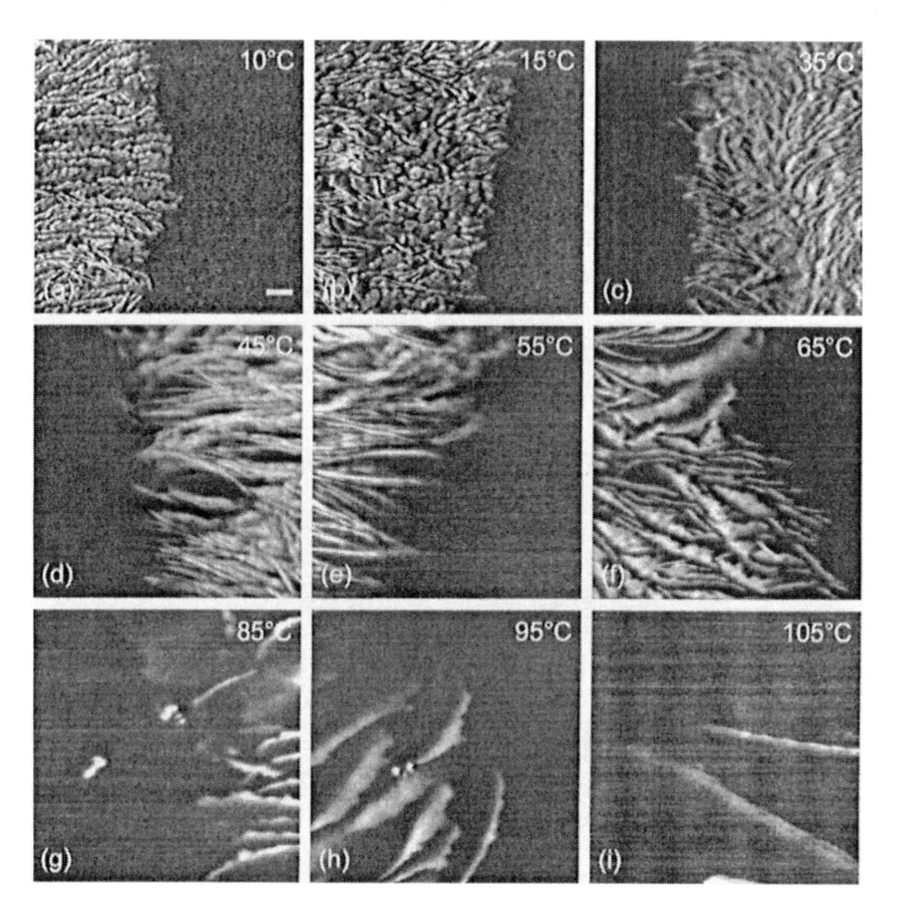

Figure 7. A series of phase images showing the lamellar growth front of spherulites of PHB/V taken at the temperatures marked on each image. Different samples were used for each measurement . (a) and (b) were taken without electronic control of the Q. (c)-(i) were taken with electronic control of the Q. (a) and (b) 32.8s/image, (c)14s/image, (d), (e), (h) and (i) 16.3 s/image, (f), (g) 11.7s/image. The scale bar represents 100nm.

We have obtained multiple images of a single lamella crystallizing over the peak in the spherulite growth rate vs. temperature curve, and over a wide range of temperatures from close to the glass transition temperature, up to close to the melting temperature. Although a detailed analysis of the new insights into polymer crystallization that can be obtained are beyond the scope of the current article, several broad observations can be made.

There is clearly no sudden change in the crystallization behaviour, but rather a gradual increase in spacing between the individual lamellae as temperature is increased. This is accompanied by an increase in distance behind the furthest forward crystallizing point at which the density of lamellae approaches its final value. Both of these observations can be explained by a reduction in the density of branching with an increase in the crystallization temperature. This behaviour will be quantified in detail elsewhere.

Examination of the edges of lamellae that are growing flat-on in the surface of the sample shows these to be rough at all temperatures. The length scale of this roughness changes with temperature, becoming progressively smaller as the crystallization temperature is reduced and the lamellae become narrower. This has implications for any nucleation theory that might be used to explain polymer crystal growth.

Ultra-high speed imaging using a resonant scanner

The electronic adjustment of the effective quality factor of the AFM cantilever has allowed us to obtain an increase of imaging speed of a factor of two or three. However, many processes of interest occur over considerably shorter timescales, and in order to construct a scanning probe microscope with a sufficiently high scan-rate to follow these processes, more fundamental limits have to be overcome.

In a new approach, described in detail in *(8)*, we have shown that it is possible to use the resonance of a microscanner as a method for mechanically positioning a probe. Here a brief outline of the technique will be given. In order to bypass the other fundamental limit to scan-rate encountered by conventional SPM, namely the time it takes to measure an interaction with the surface, we have used a near-field optical technique akin to photon scanning tunnelling microscopy (PSTM) to obtain image contrast. The probe is an etched optical fibre coated with aluminium. The sample is mounted on a glass coverslip that is then placed onto a glass prism using index matching fluid. A laser is focussed through the prism onto the sample, so as to totally internally reflect from the top surface of the glass coverslip/sample. Thus a very intense near field is formed above the sample that contains information about the local changes in optical properties of the sample. If the probe is brought into the near-field, the nanometric aperture acts to transmit this optical information to a photomultiplier (see *(8)* for more details) and thus, by scanning the probe above the surface, a high resolution optical map of the sample can be formed.

The probe is attached to a quartz crystal tuning fork with a fundamental resonance of 32kHz. The tuning fork is driven at its resonant frequency, but, rather than the amplitude being kept at several nanometres as in conventional shear-force microscopy, it is increased to several microns. Thus each resonant sweep of the cantilever traverses a line across the sample surface of several microns and forms the fast scan axis of the instrument. As the resonant frequency is 32kHz, the fast scan line rate is approximately four orders of magnitude higher than in a conventional SPM. The slow scan axis is formed by moving the microresonant scanner in a direction perpendicular to its oscillation using a conventional piezo, at a rate of up to 60Hz. Using this technique we have been able to collect a near-field optical image of part of a PHB/V spherulite in 8.3 milliseconds. Figure 8 shows an example of such an image with a schematic diagram of the experimental set-up. The optical image has a lateral resolution of between 50 and 100nm and is of a similar quality to that obtained using conventional, slow, SNOM (not shown – for an example see *(8)*).

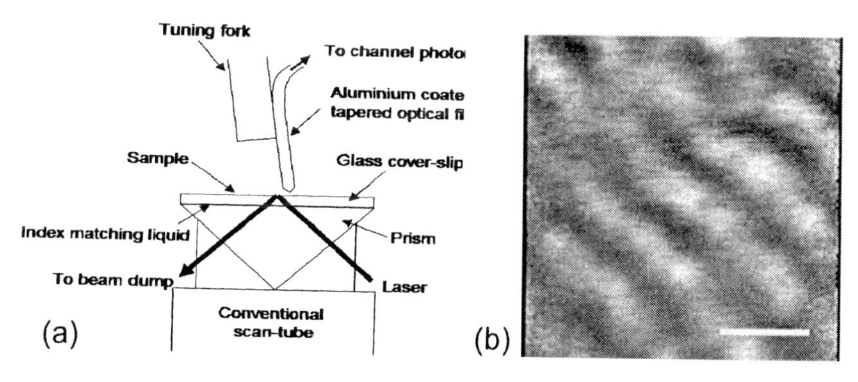

Figure 8. (a) is a schematic diagram (not to scale) showing the principle components of the resonant scanning SNOM. (b) shows a near-field optical image collected with the microscope, which is of comparative quality to those obtained imaging in the conventional manner. The scale bar represents 1 μm.

The sample probe separation is maintained at an approximately consant value by using the average optical intensity as a feedback parameter. As no adjustment of height can be carried out over the timescale of one scan-line, to a first approximation the probe addresses a plane above, and approximately parallel to, the sample surface. In fact, the motion of the probe is parabolic, but, as the resonator is very long compared to the scan area, the difference in surface probe separation due to this trajectory is only a couple of nanometres (depending on the actual scan area) and is unlikely to have a significant impact upon resolution when compared to surface roughness. If the surface is very rough, the average probe-sample separation will be quite large, as the plane mapped out by the probe must be above the highest feature, and so the average resolution will be worse than on a flatter sample. This limitation not withstanding, we have found that the aperture size and sample geometry tends to constrain resolution before the novel scanning method.

Conclusions

We have shown that, by using a relatively simple method for electronically reducing the effective settle-time of an AFM cantilever, it is possible to obtain significant increases in scan-rate that can allow new processes to be followed in-situ using AFM. However, a more fundamental re-assesment of the technology is required if more substantial increases in scan-rate are to be obtained, and we have shown one possible route in this direction, allowing imaging rates four orders of magnitude faster than in conventional SPM to be obtained.

References

1. Pearce, R.; Vancso, G.J. *Macromolecules* **1997**, *30(19)*, 5843
2. Schultz, J.M.; Miles, M.J. *J Polymer Sci. B*, **1998**, *36*, 2311
3. Hobbs, J.K.; Miles, M.J. *Macromolecules*, **2001**, *34*, 353
4. Ivanov, D.A.; Pop, T.; Yoon, D.Y.; Jonas, A.M. *Macromolecules* **2002**, *35(26)*, 9813
5. Pearce, R.; Vancso, G.J. *Polymer* **1998**, *39*, 1237
6. Neto, C.; Jacobs, K.; Seeman, R.; Blossey, R.; Becker, J.; Grun, G. *J. Phys. Cond. Mat.* **2003**, *15(1)*, S421
7. Sulchek, T.; Hsieh, R.; Adams, J.D.; Yaralioglu, G.G.; Minne, S.C.; Quate, C.F.; Cleveland, J.P.; Atalar, A.; Adderton, D.M. *App. Phys. Lett.* **2000**, *76(11)*, 1473
8. Humphris, A.D.L.; Hobbs, J.K.; Miles, M.J. *App. Phys. Lett.* **2003**, *83(1)*, 6
9. Hobbs, J.K.; Humphris, A.D.L.; Miles, M.J. *Macromolecules* **2001**, *34*, 5508
10. Humphris, A.D.L.; Tamayo, J.; Miles, M.J. *Langmuir* **2000**, *16*, 7891
11. Tamayo, J.; Humphris, A.D.L.; Miles, M.J. *App. Phys. Lett.* **2000**, *77*, 582
12. Antognozzi, M.; Szczelkun, M.D.; Humphris, A.D.L.; Miles, M.J. *App. Phys. Lett.* **2003**, *82(17)*, 2761
13. Hobbs, J.K.; McMaster, T.J.; Miles, M.J.; Barham, P.J. *Polymer*, **1998**, *39(12)*, 2437

Chapter 15

Conformation of Polymer Molecule via Atomic Force Microscopy

Sergiy Minko[1], Anton Kiriy[2], Ganna Gorodyska[2], Roman Sheparovych[1], Robert Lupitskyy[1], Constantinos Tsitsilianis[3], and Manfred Stamm[2]

[1] Department of Chemistry, Clarkson University, Potsdam, NY 13699
[2]Institute for Polymer Research at Dresden, Hohe Strossel, 01069 Dresden, Germany
[3]Department of Chemical Engineering, University of Patras, Patras, Greece

We discuss single polymer molecule experiments performed with Atomic Force Microscopy. Applying the segment-segment repulsive forces (long range electrostatic repulsion of charged units) and external mechanical forces (shear forces in a flow of liquid) polymer chains can be deposited on the solid substrate in extended coil, warm-, or rod-like conformations useful for the analysis of molecular characteristics and architecture of the polymer molecules with AFM. In the case of strong segment-substrate interactions the conformations of the deposited molecules in many details retain the solution macroscopic conformation during rapid deposition and solvent evaporation (trapped chains).

Introduction

The high resolution AFM is used to extract the information of special distribution of atoms and bonds in assembles of molecules constituting periodic structures rather than for single adsorbed molecules *(1-3)*. The study of single *polymer* molecules with AFM represents an exceptional case, when AFM can be applied due to the large dimension (length) of polymer molecules *(4,5)*. Due to the extremely high sensitivity of the AFM instruments in z-direction the polymer chain deposited on a flat surface can be easily identified even in the case of a hydrocarbon backbone *(6-9)* if the surface roughness is not larger than the size of the constituting units of the polymer chains (typically 0.2-0.5 nm). The special case is represented by giant polymer molecules such as molecular brushes or polymers with bulky side groups recently reviewed by Sheiko and Möller *(4)*.

Thus, if the contour length of the polymer chain is larger than the tip diameter (typically for molecular weight of polymer > 20 000 g/mol) the conformation of the adsorbed polymer chain can be visualized with AFM and the observed details of the conformation can be used for the study of polymer molecules. The contour length of the polymer molecule obtained from the AFM experiment can be used for the quantitative evaluation of molecular weight and molecular weight distribution *(10,11)*. Kinetic experiments (monitoring of the conformational changes by continuous scanning of the sample) allows for the study of the chain mobility on the solid surface. Most of these experiments do not require special functions and special approaches in processing and interpretation of primary AFM data.

Sample preparation is the critical point of the experiments. The conformation and behavior of the single polymer molecule which appear in the AFM image is strongly affected by the sample preparation history and the interactions with the substrate. Thus, the success of the experiments and the correct extraction of useful information from the experiments depend to a great extent on how the sample preparation modifies the conformation. From this point of view, AFM experiments with single molecules are very close to those with SEM and TEM *(12-14)*. AFM has advantages, first of all, because of the possibility to visualize the adsorbed molecule *in situ* under liquid. The potential possibility to damage/modify the conformation with a tip is very often overestimated. In many cases the effect of the tip on the observed structure can be regulated and controlled via changing the scanning conditions. However, the effect of the substrate surface and solvent evaporation (for "dry" AFM experiments) is not very predictable. The correct manipulation with those external stimuli allows for the fruitful use of AFM in the single polymer molecule experiments.

Trapped or in equilibrium state?

Polymer molecules deposited from solution onto the solid substrate may appear either in the trapped conformation or in the conformation sometimes referred to the equilibrium adsorbed state. Then the rapid evaporation of solvent can result either in the trapped conformation or the conformation which is damaged by three-dimensional (3-D) collapse or capillary forces on the border line of solvent droplets resulting in buckling and twisting of chains *(4)*. Thus, there are many steps during sample preparation when the conformation can be strongly changed by external conditions applied in the procedure.

Polymer adsorption appears in the literature as a two-step process. The first step is the diffusion-limited adsorption, when the number of adsorbed molecules increases with the square root of time *(15)*. Chains enter into contact with the surface and adsorb, retaining their solution conformation. The second step is slow reconformation of the adsorbed chains, when the chains become progressively flatter *(16)*. Changes occurring during a long period of time of equilibration in adsorbed layers were found to be very slow and have a complicated, nonequilibrium character *(17-21)*. Reconformation characteristic time of the adsorbed chains differs from seconds to hours depending on interactions in the particular system. The polymer molecule can be kinetically trapped by the substrate, and changes in conformation can be detected after a long period of time. This particular case is of special interest for investigations of conformations in solution. In this case, the polymer chain is adsorbed on the substrate surface, retaining its solution conformation in many details, and its two-dimensional (2-D) size correlates with the dimensions of the chain in solvent.

We have found conditions providing this particular case of frozen chain conformations, when the dimension of polymer molecule in solution is very close to this value for polyelectrolyte (PE) molecules deposited on mica substrate *(22-24)*. PE molecules are of special interest for AFM single molecule experiments because of the extended conformation of the chains affected by electrostatic repulsion between charged monomer units. The charge density and the electrostatic interactions can be tuned by the change of pH and ionic strength resulting in the change of conformations. Thus, PE are attractive objects for AFM experiments.

In Figure 1, we outlined the experimental illustration of two different cases of a change of the molecular conformation of poly(methacryloyloxyethyl dimethylbenzylammonium chloride) (PMB, M_n= 480 kg/mol; M_w=720 kg/mol) chains adsorbed on mica substrate. PMB is highly charged polyelectrolyte (polycation) in aqueous solution that possesses the extended coil or worm-like conformation. For simplicity, we characterize the size of the adsorbed coils by the value L which is measured as a long length axis of the structures visualized

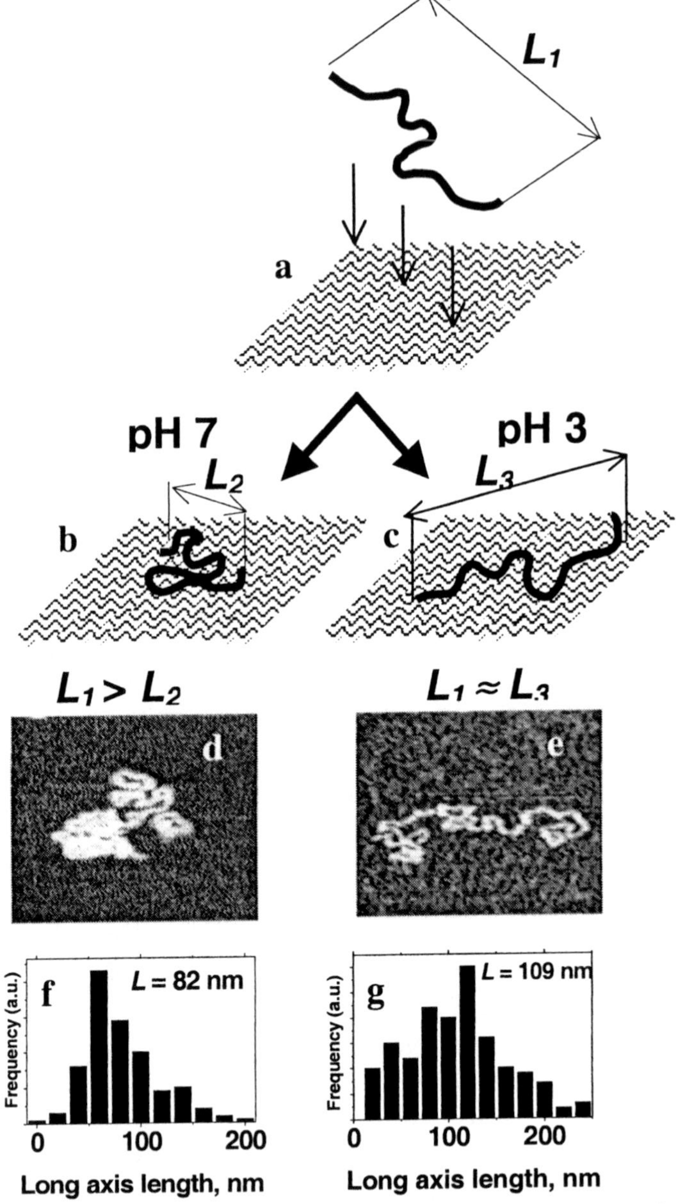

Figure 1. Adsorption of PMB molecules on solid substrate: outline of two possible cases when molecule (a) with the characteristic size L adsorbed and trapped with dramatic change of conformation (b) or with very small change of dimensions (c) and experimental representative AFM images of the conformations on negatively charged (potential -28 mV) (d) and at IEP of mica (e), respectively; histograms showing the differences of the average L after adsorption for both cases (f) and (g), respectively.
(Reproduced from reference 22. Copyright 2002 American Chemical Society.)

with AFM *(22)*. The PMB chain of characteristic dimension L_1 approaches the mica surface and appears in the dry state as a wormlike coil with the size L_2 or L_3 when adsorbed at pH 3 and pH 7, respectively. Figure 1d and e show representative AFM images of the adsorbed chains, while in Figure 1f and g we present corresponding histograms obtained from measurements of 150 molecules (from about 15 different images). We may conclude that $L_2 < L_3$. At pH 3, the mica surface is only slightly negatively charged, while a much larger negative charge at pH 7 enhances screening of the intrachain Coulombic repulsion in PMB, which appears in the more coiled conformation in good agreement with the theoretical prediction of adsorption of hydrophobic polyelectrolytes on oppositely charged surfaces *(25)*. Thus, we performed the experiments near the IEP of mica at a pH ranging from 2 to 3, assuming that in this case the surface introduces minimal changes in PE molecular conformations.

We found very good correlation between L_3 and L_1 values comparing results of light-scattering experiments with AFM data for PE molecules of different length and for different ionic strengths *(22)* of PE solutions which gave evidence that the molecular conformation was not substantially changed during adsorption and following drying.

Similar experiments we performed with the star-like block-copolymers constituted of 7 polystyrene (PS) (Mw=20 kg/mol) and 7 poly(2-vinylpyridine) (P2VP) (Mw =56.5 kg/mol) arms (PS_7-$P2VP_7$) emanated from the same core

Table 1. Comparison of the Molecular Radius (*R*) of PS_7-$P2VP_7$ Heteroarm Star Copolymer in Solution and onto the Mica Surface after the Rapid Deposition

	radius, nm		
solvent	dynamic light scattering, R_h	gel permeation chromatography R_{GPC}	AFM R_{AFM}
THF		15	18 ± 5
toluene	31		40 ± 5
water, pH 2 (micelles)	85		97 ± 20

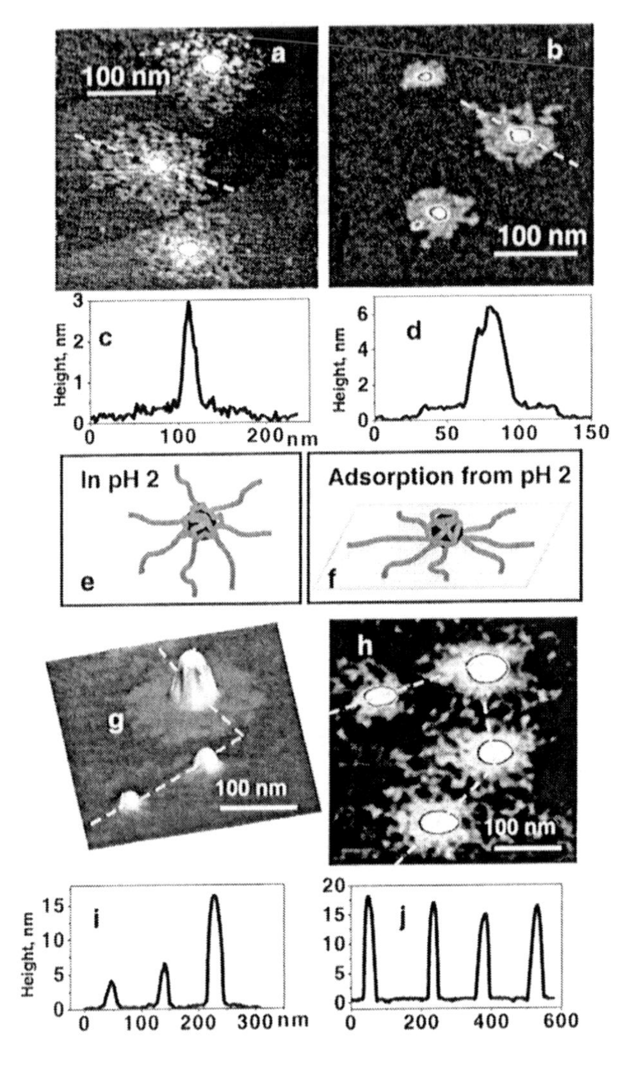

Figure 2. AFM topographic images (a, b, g, h) and cross sections (c, d, i, j) of PS_7-$P2VP_7$ adsorbed from acid water (pH 2, HCl): in salt-free condition (a,c) and in the presence of 1 mM of Na_3PO_4 (b,d). Schematic representation of the unimolecular micelle formed in acid water: in solution (e); in adsorbed state (f). 3D image (g) and cross section (i) of micelles with different aggregation number formed at pH 3.5 AFM images (h) and cross section (j) of multimolecular micelles adsorbed from acid water (pH 4.2, HCl) (Reproduced from reference 24. Copyright 2003 American Chemical Society.)

(Figure 2). Again, we study the rapidly deposited star-like molecules *(24)*. In Table 1 we compare sizes of the single molecules evaluated with different methods in solution with the size of the outer shell of the molecules obtained in the AFM experiments. The quite good agreement demonstrates that in this case the change of conformation on the surface is slow and it is not strongly changed during the deposition time.

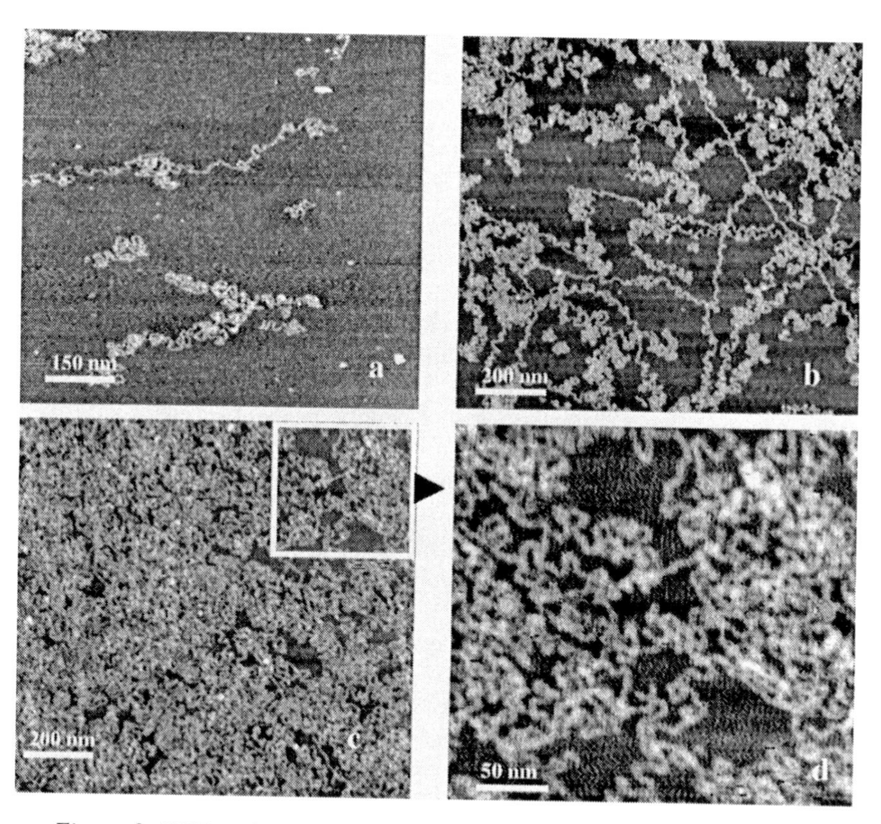

Figure 3. PMB molecules adsorbed on the mica from salt-free solution at polymer concentration 0.01 g/l (a,b) and 0.02 g/l (c,d) for 5 second (a); 30 second (b); and 3 minutes (c, d).

Strongly opposite results were obtained for the case of polyanions. Sodium salts of polyacrylic acid and polystyrenesulfonic acids deposited on mica in similar experiments were observed in hypercoiled confoiormations. The interaction with the substrate was not enough strong to trap the molecules in extended conformations. Even in the case where the chains were stretched by shear forces in a flow of water and dried we observed a fast transition from the extended into hypercoiled conformations upon a short exposure of the sample to water.

Study of adsorption

AFM experiments for study of polymer adsorption can be extended to the assemblies of molecules when the adsorption process runs until many single molecules and the groups of the molecules on the surface are visualized. At a longer time or at a larger concentration of the molecules on the surface they start to overlap and interact with each other. In Figure 3 we present the example of the adsorption of PMB molecules on mica surface for 3 min. The images allow us to observe the conformations of individual chains until the some limiting surface concentration is approached. At this concentration because of the poor lateral resolution we can not follow an individual polymer chain in the assembly. However, the images help to study conformations in the concentrated adsorbed layer.

Reconstruction of the conformation in solution

Conformations of the trapped polymer molecules reflect to some extent the conformation in solution. Although, the tiny details of the conformation are lost due to the adsorption and solvent evaporation, however some important details of the macroconformation may be extracted from the investigation of the trapped polymer chains. For example, the coil-globular transition in solvent can be very precisely investigated with AFM experiments *(4)*. This transition is usually observed as dramatic conformational changes when the conformation switches from a coil-, worm-, or rod-like chain conformation to the compact globule conformation when dimensions of the chains change in 10-100 folds. Such substantial conformational alterations are frozen and observed in the trapped chains.

In Figure 4, we present the series of experiments when we study the stepwise coil to globule transition (CGT) of PMB molecules in aqueous solutions by adding Na_3PO_4 *(22)*. In salt-free solutions, the polymer molecules appear as extended coils (Figure 4a). Added salt dramatically changes molecular conformations and the fine morphology of the PE chains. Images in Figure 4b-n clearly show intramolecular segregated areas. Two pronounced differences from salt-free solution can be found on the images: the polymer coil starts to segregate into small beads nicely observed in the zoom image in Figure 4e,g; the beads segregate into big clusters. The height of the beads is several times larger than the height of the backbone. We assume according to the DRO model *(26)* that

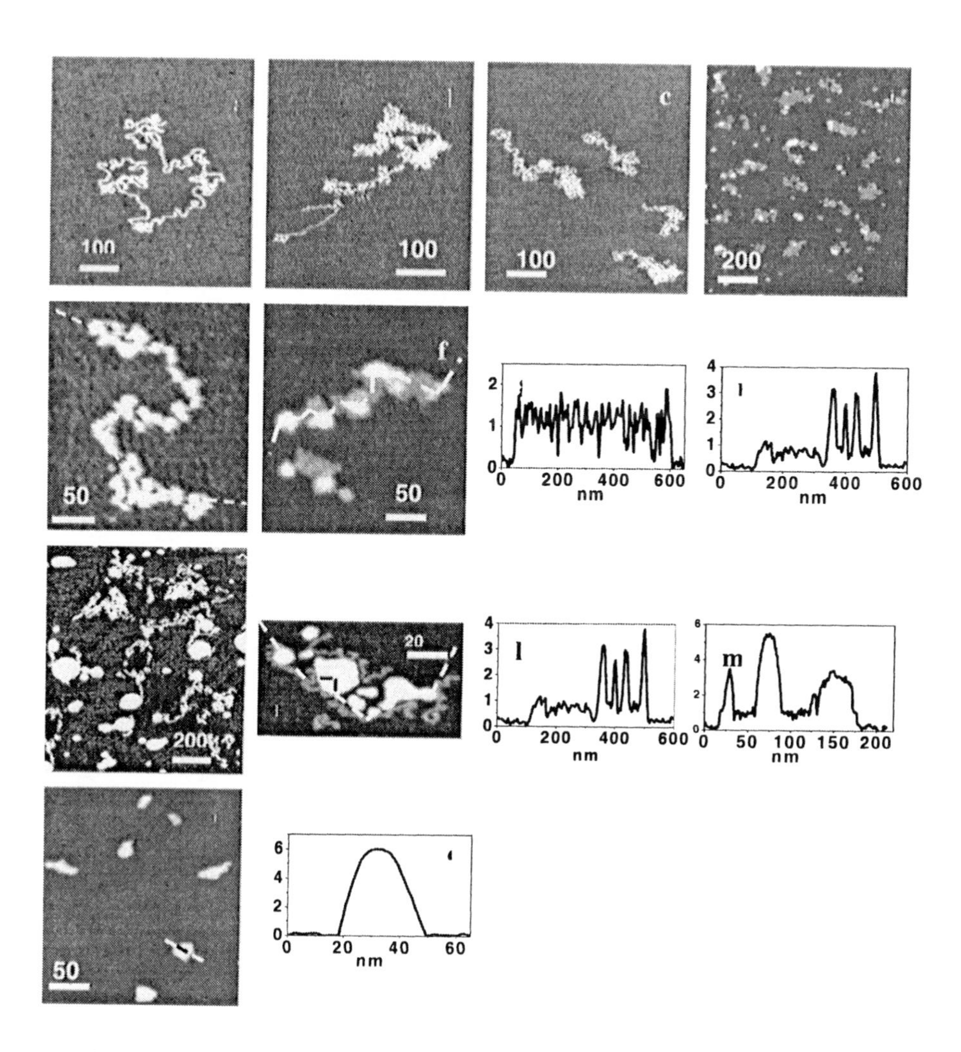

Figure 4. AFM images of PMB single molecules deposited from aqueous solutions: reference, no salt (a) and with added Na_3PO_4: 4.2 nM (b); 6 mM (c and e); 8.4 mM (d and f); 8.7 mM (i); 12 mM (j); 18 mM (m). Cross sections from AFM images: (g) corresponds to (e); (h) corresponds to (f); (k) corresponds to (i); (l) corresponds to (j); (n) corresponds to (m). (Reproduced from reference 22. Copyright 2002 American Chemical Society.)

the beads are formed due to the PE intrachain segregation induced by the screening effect of the added salt and due to the enhanced attraction induced by condensed salt ions. Each step of adding salt results in the increase of the size of beads and the decrease of their number, as well as the decrease of the necklace length.

The changes of macroconformations in solution can be successfully reconstructed and used for the investigations of the polymer molecules of a complicated architecture. The experiments with heteroarm star-like block-copolymer PS_7-$P2VP_7$ give the convinced evidence for that *(24,27)*.

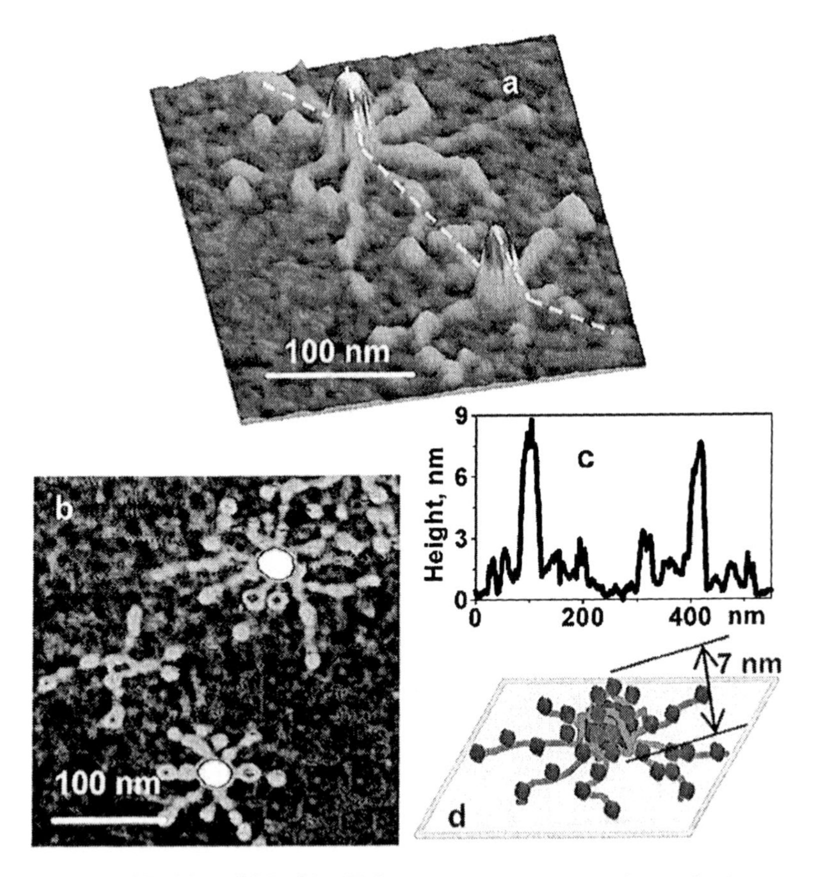

Figure 5. 3D (a) and 2D (b) AFM images, cross section (c), and schematic representation (d) of Pd-metallized unimers. (Reproduced from reference 24. Copyright 2003 American Chemical Society.)

It was previously shown that at a relatively high concentration (0.3 g/L) PS_7-$P2VP_7$ undergoes intermolecular micellization in acid water (pH 1-2) with the aggregation number equal to 8 *(28)*. In contrast, at extremely low concentrations

(0.005 g/L of PS_7-$P2VP_7$) micellization occurs as an intramolecular process. In these conditions the PS_7-$P2VP_7$ star copolymer survives in the nonassociated state and forms stable unimolecular micelles. Figure 2 shows representative AFM images of mica-deposited uniform star-shaped core-shell structures formed upon dilution of PS_7-$P2VP_7$/THF solution with acid water (pH 2). Such a morphology reflects very pronounced intrasegregation of the star copolymer in acid water *(29)*. PS chains collapse due to hydrophobic interactions and form a compact core whereas protonated P2VP arms adopt an extended conformation due to the Coulomb repulsion and form a shell (Figure 2e). We may conclude that the AFM image represents an "off print" of the solution conformation when the molecules were adsorbed and trapped by the substrate. After the rapid evaporation of water the structure was formed due to the collapse in the *Z*-direction (Figure 2f). The experimental value of the collapsed PS core volume is about 219 nm^3, which corresponds to the calculated value (245 nm^3).

To improve AFM resolution of P2VP arms, we employed a recently developed approach to decorate polyelectrolyte chains with metal clusters. This method consists of the ion-exchange reaction between protonated $(P2VPH^+)Cl^-$ and Na_2PdCl_4 and the following reduction of Pd with dimethylamine borane *(11,24,27,30,31)*. After metallization PS_7-$P2VP_7$ deposited on the Si wafer from acidic aqueous solution with fully extended P2VP arms appears in very good resolved star-shaped conformations (Figure 5). Metallization of the adsorbed PS_7-$P2VP_7$ molecules allowed us to get better insight into the structure of the unimolecular micelles. Specifically, we found the considerable increase of the height of the cores occurred upon metallization (from H = 2.9 nm before metallization to H = 7-12 nm after). Ion-exchange reaction of palladium tetrachloride dianion serves a selective and efficient probe on polycations, and thus, such an observation clearly indicates that pyridine units are localized in the shell and in the core of the micelles (Figure 5d). This contrasting method will be discussed below.

It is well documented that the critical micelle concentration (cmc) of similarly designed PS_6-$P2VP_6$ copolymers in toluene is near 0.7 mg/mL and that at concentrations below cmc PS_6-$P2VP_6$ forms inverse unimolecular micelles *(32)*. AFM experiments confirm that for PS_7-$P2VP_7$ immediately after addition of toluene (good solvent for PS arms and bad solvent for P2VP arms) into the PS_7-$P2VP_7$/THF solution, intramolecular segregation occurs rather than aggregation. Unimers initially formed upon addition of toluene are poorly segregated particles consisting of swollen cores ($D = 44 \pm 5$ nm) and small shells ($D = 55 \pm 5$ nm). A full reconformation of the unimers occurs within 1 h. Star-like structures with well-definite PS arms (about 30 nm in length) in the shell and P2VP arms collapsed in the core were observed *(24)*. The core diameter is $D = 36 \pm 10$ nm, about twice of the PS core that is observed for the unimolecular micelles in water. The latter is consistent with the composition of the copolymer

($W_{P2PV}/W_{PS} \approx 2$). Moreover, the P2VP core volume calculated from AFM data is V_{AFM} = 813 nm^3, which is slightly higher than the volume of P2VP blocks calculated from the degree of polymerization ($V_{P2VP,calc}$ = 698 nm^3). That provides evidence for the formation of the inverse unimolecular micelles upon addition of toluene.

Intermolecular interactions

Chain-chain interactions in solution resulting in a strong alteration of the conformation or in an association of polymer chains can be investigated with AFM using the same concept. We assume that the rapid deposition and solvent evaporation freezes the macroconformations of the interacting polymer chains and structures of the aggregates.

Unimolecular micelles PS_7-$P2VP_7$ were found to be stable in dilute solutions at pH 2 for at least several days. At pH 3.5 about 20% of PS_7-$P2VP_7$ molecules undergo association, leading to multimolecular micelles, whereas most of molecules remain unassociated (unimers) with slightly compressed shells. Figure 2 shows clearly the coexistence of micelles and unimers with significant differences on their size. At pH 4.2, the association phenomena lead to multimolecular micelles (Figure 2h,j). The size of the PS cores ($H = 17 \pm 1$; $D = 22 \pm 5$) has been increased significantly as compared to unimolecular micelles. The association is also reflected in an increase of the diameter of the shell ($D = 137 \pm 35$) if more P2VP chains are accumulated in the corona, resulting in larger stretching due to the excluded-volume effect. The formation of micelles at pH 4.2 (Figure 2h,j) may be attributed to the fact that a considerable part of P2VP exists in nonprotonated state, altering the hydrophobic-hydrophilic balance of the copolymer and therefore destabilizing the unimers, inducing association.

Chemical contrasting

Single polymer molecules were visualized on an atomically flat mica surface *(9,22-24)*. Although these PE are invisible on a Si-wafer due to the high roughness of the substrate, they can be resolved after decoration of the deposited chains with Pd clusters *(11,27,31)*. Metallization of PE causes the strong contraction of chains (2-3 times decrease of the contour length) even if the chains were strongly trapped by the substrate. The local collapse of the chain was induced by interaction with the bivalent ion $PdCl_4^{2-}$.

We developed a contrasting procedure which allows us to improve substantially the resolution of a single molecule experiment with no changes of the conformation of adsorbed polymer molecules *(33)*. In our approach, we use the deposition of either hexacyanoferrate (HCF) anions or negatively charged clusters of cyanide-bridged complexes as contrasting agents (Figure 6). This method allowed us to increase the thickness of the resulting structures up to 3 nm and, consequently, to provide visualization of polymer chains on Si-wafers. After AFM-measurements, the contrasting agents were then removed also without distortion of the molecule conformation.

For example, both PMB and P2VP molecules deposited onto the Si-wafer are not resolved in the tapping mode (Figure 6b). Although PS_7-$P2VP_7$ molecules adsorbed onto mica from acid water (pH 2) solution display a clear core-shell morphology (Figure 6e), only the core of unimers with the height of 5 nm can be resolved on the Si-wafer (Figure 6f). We were successful in visualizing all of these polymers on the Si-wafer substrate with the following staining procedure. The polymer molecules were deposited on the freshly cleaved mica or Si-wafers from a very diluted (0.0005 mg/mL) acid solution (pH 2.5-3). The drop of the examining solution was set on the substrate for 60 s, and afterward it was removed with a centrifugal force. The samples were then stained upon exposure for 3 min to an HCF acid solution bath. The sample was rinsed in water and dried for the AFM experiment. Figure 6c,g presents AFM images of PMB and PS_7-$P2VP_7$ molecules contrasted with HCF. In all cases, we observed an 0.6-0.7 nm increase of heights of resulting structures, that roughly corresponded to the size of the HCF-anion (Figure 6a). We found a strong effect of pH on the contrasting process. No attachment of HCF-anions was observed at pH higher than 4, that is, above the isoelectric point of the Si-wafer (pH 3.8). We may speculate that at the large pH the negatively charged Si-wafer suppresses the interaction of the polycations with HCF-anions. The statistical analysis of the molecular diameter from AFM images of the PS_7-$P2VP_7$ unimers before the contrasting on mica and after the contrasting on the Si-wafer provides evidence that the contrasting procedure introduces no changes in the conformation (Figure 6h,j). Similarly, we detected no changes of the dimensions of PMB molecules upon staining with HCF. We found that deposited HCF can be removed without changes of the molecular dimensions of PE simply upon rinsing the sample with either acidic (HCl, 5%) or basic (NH_3, 3%) water solution for several minutes.

Molecular Characteristics

AFM single molecule experiment can be employed as a direct analytical method to estimate molecular characteristics (constitution) of polymer molecules such as

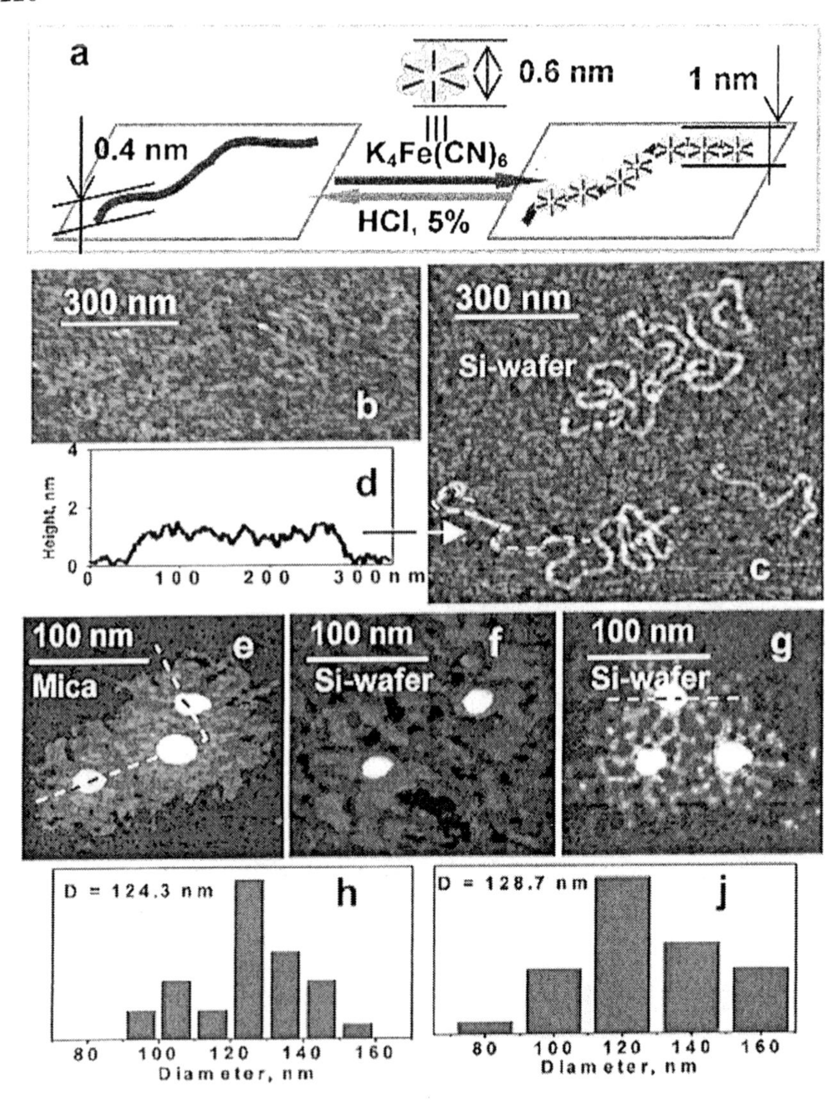

Figure 6. Scheme of the contrasting of adsorbed polycations (a), AFM topography images (b,c), and a cross-section (d) of PMB, and AFM images of PS_7-P2VP (e-g) molecules before (b,e,f) and after (c,g) contrasting with HCF (Z-range 5 nm). All images are on Si-wafers, but (e) is on mica. Histograms of molecular diameter distribution for PS_7-$PsVP_7$ adsorbed onto the mica, with no contrasting, are shown in (h), and those onto the Si-wafer after contrasting with HCF are shown in (j). (Reproduced from reference 33. Copyright 2003 American Chemical Society.)

a contour length, molecular weight, degree of branching. This option is of special importannce in the case of branched macromolecules with a complicated architecture (block-copolymers, star-like copolymers, molecular brushes, etc). when traditional methods (light scattering, GPC) can not deliver reliable or complete data on the molecular characteristics. Recently, AFM was successfully used for the *quantitative* analysis of polymer molecules *(10,22-24)*.

For example, we have evaluated the dimensions of PMB molecules *(22)*: number average l_n= 319 nm and weight average l_w= 454 nm contour length, and polydispersity index l_w/l_n = 1.4 obtained from the statistical analysis of 150 structures on 18 AFM images. The estimated from these data values for number average molecular weight M_n= 360 000 g/mol and weight average molecular weight M_w= 512 000 g/mol, and polydispersity index PD =1.5 were very close to the results of traditional methods. Static and dynamic LS, ultracentrifugation method, and GPS have shown quite strong scattering of the data between different methods (M_n= 480 000 g/mol, M_w= 720 000 g/mol, PD= 1.4) which is a typical situation for large charged polymer molecules in aqueous solution.

In the other example we investigate molecular architecture of the PS_7-$P2VP_7$ molecules *(24)*. The P2VP arms decorated with Pd clusters or contrasted with HCF clusters are clearly observed in the AFM images and can be counted. The AFM images for the first time visualized the second generation of the P2VP arms, which were growing up from the active sites located on the core of the PS star polymer precursor. Although the first generation of the PS arms was characterized (number of arms) by light scattering from the molecular weights of the star copolymer and the arms, the characterization of the P2VP arms was performed only due to the visualization. The average number of P2VP arms was counted directly from the AFM images to be 7 ± 1.26, which was in excellent agreement with the number of PS arms found by light scattering. This result gives unambiguous evidence that the number of the chemically different arms is equal. At the same time, that is very exciting example how AFM visualization can be used for analysis of polymer architecture.

Conclusions

Our brief analysis of recent experiments with single polymer molecules and assemblies using AFM visualization delivers convincing arguments for the fruitful application of this approach in polymer science. The initial skepticism based on the doubts that deposited polymer molecules appear in unpredictable conformations affected by numerous parameters of the polymer adsorption and solvent evaporation have to be exchanged with the understanding of the role of the sample history. Yes, the sample history effects the conformation. However, that is exactly the strongest side of the experiments. Our results show that quite important information about polymer conformations can be recovered using the sample history. The further extension of the method requires analysis of the external stimuli affected conformational changes of the deposited polymer molecules as well as a comparison of the conformations in solution and on the surface at different conditions of the sample preparation and treatment.

References

1. Magonov, S. N.; Whangbo, M. H. *Surface analysis with STM and AFM*; VCH: Weinheim, 1996.
2. Magonov, S. N. In *Encyclopedia of Analytical Chemistry*; Meyers, R. A., Ed.; Wiley & Sons: Chichester, U.K., 2000; pp 7432 - 7491.
3. Chernoff, D. A.; Magonov, S. N. In *Comprehensive desk reference of polymer characterization and analysis*; Brady, R. F., Ed.; Oxford University Press: Washington, D.C., 2003; pp 490 - 531.
4. Sheiko, S. S.; Moller, M. *Chem Rev* **2001**, *101*, 4099-4123.
5. Hashimoto, T.; Okumura, A.; Tanabe, D. *Macromolecules* **2003**, *36*, 7324-7330.
6. Kumaki, J.; Nishikawa, Y.; Hashimoto, T. *J. Am. Chem. Soc.* **1996**, *118*, 3321-3322.
7. Sakurai, S.; Kuroyanagi, K.; Morino, K.; Kunitake, M.; Yashima, E. *Macromolecules* **2003**, *36*, 9670-9674.
8. Kumaki, J.; Hashimoto, T. *J. Am. Chem. Soc.* **2003**, *125*, 4907-4917.
9. Minko, S.; Gorodyska, G.; Kiriy, A.; Jaeger, W.; Stamm, M. *Abstr Pap Am Chem S* **2002**, *224*, U529-U529.
10. Sheiko, S. S.; da Silva, M.; Shirvaniants, D.; LaRue, I.; Prokhorova, S.; Moeller, M.; Beers, K.; Matyjaszewski, K. *J Am Chem Soc* **2003**, *125*, 6725-6728.
11. Minko, S.; Kiriy, A.; Gorodyska, G.; Stamm, M. *J Am Chem Soc* **2002**, *124*, 10192-10197.
12. Zhang, M.; Drechsler, M.; Muller, A. H. E. *Chem. Mater.* **2004**, *16*, 537-543.
13. Lei, L.; Gohy, J.-F.; Willet, N.; Zhang, J.-X.; Varshney, S.; Jerome, R. *Macromolecules* **2004**, *37*, 1089-1094.
14. Kirwan, L. J.; Papastavrou, G.; Borkovec, M.; Behrens, S. H. *Nano Lett* **2004**, *4*, 149-152.

15. Motschmann, H.; Stamm, M.; Toprakcioglu, C. *Macromolecules* **1991**, *24*, 3681.
16. Franz, P.; Granick, S. *Phys. Rev. Lett.*, **1991**, *66*, 899.
17. Pefferekorn, E.; Carroy, R.; Varoqui, R. *J. Polym. Sci., Polym. Phys. Ed* **1985**, *23*, 1997.
18. Pefferkorn, E.; Haouam, A.; Varoqui, R. *Macromolecules* **1989**, *22*, 2667.
19. Johnson, H. E.; Granick, S. *Macromolecules* **1990**, *23*, 3367.
20. Johnson, H. E.; Granick, S. *Science* **1992**, *255*, 966.
21. Voronov, A.; Pefferkorn, E.; Minko, S. *Macromol. Rapid Commun.* **1999**, *20*, 85-87.
22. Kiriy, A.; Gorodyska, G.; Minko, S.; Jaeger, W.; Stepanek, P.; Stamm, M. *J Am Chem Soc* **2002**, *124*, 13454-13462.
23. Minko, S.; Kiriy, A.; Gorodyska, G.; Stamm, M. *J Am Chem Soc* **2002**, *124*, 3218-3219.
24. Kiriy, A.; Gorodyska, G.; Minko, S.; Stamm, M.; Tsitsilianis, C. *Macromolecules* **2003**, *36*, 8704-8711.
25. Borisov, O. V.; Hake, F.; Vilgis, T. A.; Joanny, J.-F.; Johner, A. *Eur. Phys. J.* **2001**, *6*, 37-47.
26. Dobrynin, A. V.; Rubinstein, M.; Obukhov, S. P. *Macromolecules* **1996**, *29*, 2974-2979.
27. Gorodyska, G.; Kiriy, A.; Minko, S.; Tsitsilianis, C.; Stamm, M. *Nano Lett* **2003**, *3*, 365-368.
28. Tsitsilianis, C.; Voulgaris, D.; Stepanek, M.; Podhajecka, K.; Prochazka, K.; Tuzar, Z.; Brown, W. *Langmuir* **2000**, *16*, 6868.
29. Wolterink, J.; Leermakers, F.; Fleer, G.; Koopal, L.; Zhulina, E.; Borisov, O. *Macromolecules* **1999**, *32*, 2365.
30. Kiriy, A.; Minko, S.; Gorodyska, G.; Stamm, M.; Jaeger, W. *Nano Lett* **2002**, *2*, 881-885.
31. Minko, S.; Gorodyska, G.; Kiriy, A.; Stamm, M. *Abstr Pap Am Chem S* **2002**, *224*, U402-U402.
32. Voulgaris, D.; Tsitsilianis, C.; Esselink, F.; Hadziioannou, G. *Polymer* **1998**, *39*, 6429.
33. Kiriy, A.; Gorodyska, G.; Minko, S.; Tsitsilianis, C.; Jaeger, W.; Stamm, M. *J Am Chem Soc* **2003**, *125*, 11202-11203.

Chapter 16

Solving Practical Problems in the Plastics Industry with Atomic Force Mircroscopy

Francis M. Mirabella, Jr.

Dilute Solution Properties and Polymer Physcis Department, Equistar Technology Center, 11530 Northlake Drive Cincinnati, OH 45249

Introduction

The plastics industry has undergone revolutionary changes, since the first applications of long-chain molecules as commercial materials. The driving force to the development of plastics as an alternative to metal, wood, glass, natural rubber and paper has been the invention of complex systems, which rival or surpass these other materials in strength/weight ratio, stiffness, toughness, clarity, etc. The characterization of complex polymer systems involves microstructure analysis such as molecular weight, copolymer composition, stereoregularity, monomer sequence distribution, etc. Microstructural variables are determined by such methods as nuclear magnetic resonance (NMR), infrared spectroscopy (IR), differential scanning calorimetry (DSC), size exclusion chromatography (SEC), etc.

Although microstructural variables are a useful component of the information required for predicting the properties of complex polymer systems, they are insufficient information for predicting physical and mechanical properties. It is necessary to characterize the solid-state morphology of such systems in order to accurately predict these properties, since the morphological structure has a profound effect on the micro-mechanical behavior of polymer systems. Atomic force microscopy (AFM) has distinguished itself as a central technique for the characterization of polymers[1], especially multi-phase polymer systems.

In these studies both topographic and phase-contrast data were obtained from the AFM in order to infer the phase-structure information. The phase-structure information developed from AFM data was used to explain the mechanical behavior of these systems. Detailed discussion of the AFM operational modes, origin of and instrumental basis for topographic and phase imaging, practical considerations, etc. can be found in the review by Chernov and Magonov.[1]

Experimental

AFM

A Digital Instruments (Veeco Instruments, Inc., Santa Barbara, CA) NanoScope® IIIA/Dimension 3000 large sample AFM, with phase extender was operated in air using TappingMode™ to capture height and phase images.

Sample Preparation

Pellet samples were compression molded and quenched in cold tap water. Plaques were cut from the injection moldings to fit into the ultra-microtome chuck. The natural and microtomed surfaces observed in the AFM were rinsed with methanol and dried in a compressed air stream to remove surface debris, oils, dirt, etc.

Microtomy

Compression molded samples were cut in a Leica Utracut UCT® ultra-microtome with a EMFCS™ cryo-stage (Leica Mikrosysteme GmbH, Vienna, Austria). An initial cut was made with a glass knife in order to produce a flat face on the sample. Subsequently a diamond knife was used to microtome the finished smooth face on the sample. Samples were held at cryogenic temperatures of -50°C to -90°C during microtoming, depending on sample brittleness. AFM was done on this smooth face.

Comparison of Techniques

The electron microscopy techniques, scanning electron microscopy (SEM) and transmission electon microscopy (TEM), have dominated the high magnification and high resolution characterization of polymer systems for five decades.[2] A comparison of the electron microscopy techniques to AFM will be discussed in this section in order to point out some of the advantages of AFM over other techniques. A comparison of sample preparation and observational differences in SEM and TEM to AFM will be given.

SEM

Preparation of SEM samples is tedious. SEM requires microtoming of the sample to reveal a flat surface. Gold coating of the surface is always required for polymers to prevent charging of the non-conducting polymer surface. If "phase" information (phase information includes distinguishing of crystalline, amorphous structure and domain structure in multi-phase polymers, etc.) is desired, etching of the sample is required to differentiate domains. There are several problems which are often associated with these and other aspects of sample preparation for and operation of SEM. Etching can cause surface modification, beyond that desired to reveal phase information. The degree of etching is uncertain, sometimes requiring multiple etching trials (e.g. solvent choice, time, temperature and agitation method, such as sonication). If the etching is incomplete the image may be corrupted.[2] Etching typically destroys internal structure, e.g. crystalline material, which resides inside etchable domains, e.g. rubbery domains dispersed in a hard, polymer matrix[2]. Although SEM samples are gold coated to minimize charging, the occurrence of "halos", due to charging, around certain domains, such as depressions of relatively large depth in the surface, causes great difficulty in the measurement of particle dimensions.[2] This problem is often the case with samples in which soft domains have been etched out of a hard matrix. Topographic features are often not observed in SEM, especially if these are small in height. Thin films often can burn through, even when gold coated.

TEM

Preparation of TEM samples is extremely tedious. Sections typically must be microtomed, which are uniform and in the thickness range of about 20-40 nm. The cutting, collecting, transfer and mounting of such sections is extremely tedious and requires a high degree of technical expertise. If features are to be distinguished, staining with a heavy metal such as Ru or Os is required. Such staining is uncertain, as to level of contrast, and often must be repeated several times to reach desired contrast. Also, staining with heavy metals is tedious and dangerous. Further, staining may cause surface modifications, which are not anticipated or detected.

AFM

AFM has many advantages, related to sample preparation, over SEM and TEM. AFM typically requires no sample preparation or only microtoming of the sample to reveal a flat surface. The simple rinsing of the surface, e.g. with methanol and drying with compressed air, often dramatically improves the image quality by the removal of surface debris collected on the sample during microtomy.[2] Rinsing of natural surfaces, which have undergone no sample preparation, is also effective for removing debris, oils, etc. One of the most useful applications of AFM is to the characterization of the phase-structure of multi-phase polymer systems. For example, the domain structure of a high impact polypropylene (IP) and thermoplastic olefins (TPO) may be readily observed in samples prepared for AFM by simply microtoming a flat surface, or in some cases, by observing the natural surface. The morphology of a TPO is shown in Figure 1. The composition of the TPO was about 75%wt. of isotactic polypropylene (PP) and 25%wt, of ethylene/propylene rubber (EPR). The EPR typically contains about 20-30%wt. of ethylene, which yields a rubber with essentially zero crystallinity. The soft, "rubbery" domains, composed of the

EPR, in the hard, crystalline isotactic polypropylene matrix may be clearly observed in Figure 1. Further, the internal structure of the rubbery domains, which are etched out (and, therefore, destroyed) for observation by SEM, is normally preserved in samples prepared for AFM. That is, crystalline domains inside the dispersed "rubbery" domains in the TPO are readily observed in the phase image in Figure 1. The crystalline domains inside the rubbery EPR particles are known to be composed of crystalline polyethylene (PE), which resides inside the EPR domains due to the phase thermodynamics operational in these systems.[3]

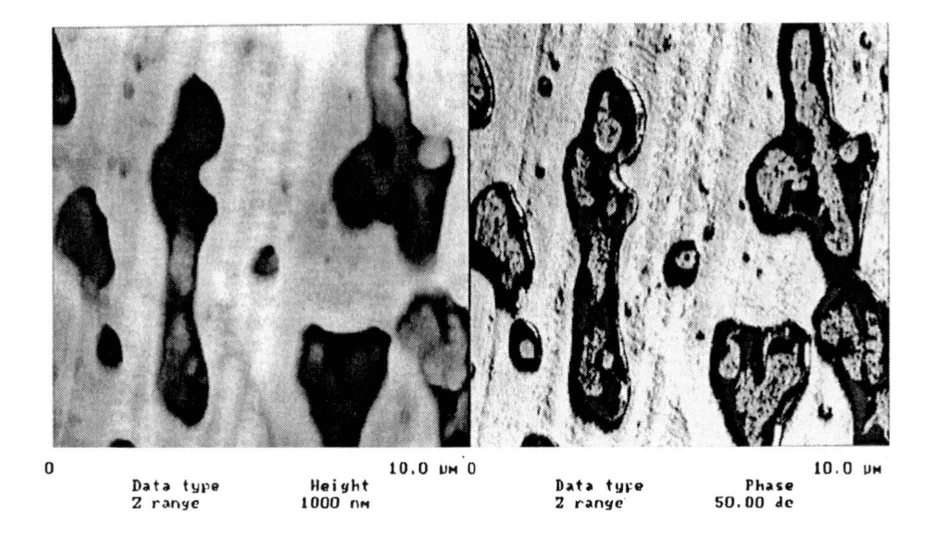

Figure 1 – AFM topographic (left image) and phase contrast (right image) photomicrographs of a TPO (thermoplastic olefin). The sample was compression molded and a flat face microtomed for AFM observation. Reproduced with permission from reference 2. Copyright 2005 Taylor and Francis Group.

Figure 2 shows three AFM phase contrast micrographs of another TPO. The lowest magnification (top) image shows the overall phase structure of the phase-separated system in which the lighter areas are crystalline and the darker areas are noncrystalline. The center image shows a detail of several noncrystalline EPR particles surrounded by the crystalline PP matrix and PE crystalline material inside the EPR particle. The highest magnification (bottom) image shows a close-up view of the interface between the PP matrix and an EPR particle with PE crystalline material in the central area of the EPR particle.

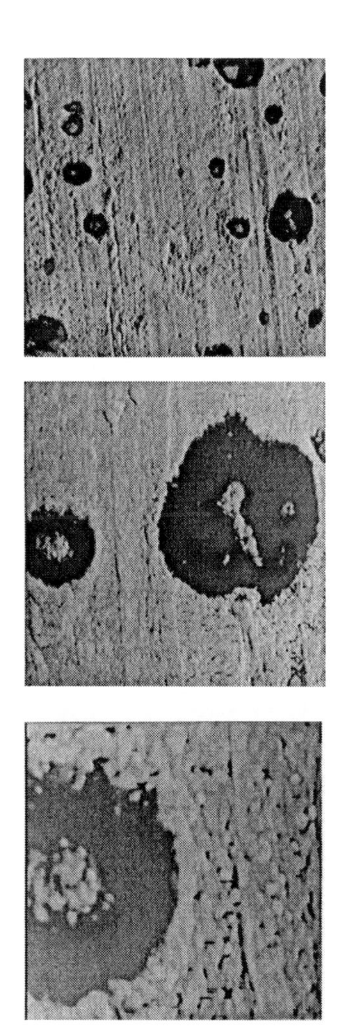

Figure 2 – AFM phase contrast photomicrographs of a TPO (thermoplastic olefin). Top image 10μm, center image 3μm and bottom image 1μm square. The sample was compression molded and a flat face microtomed for AFM observation.

Morphology/Mechanical Property Correlations in Multiphase Polyolefin

Atomic force microscopy was used to characterize a series of multi-phase impact polypropylene (IP)/high density polyethylene (HDPE) blends. The composition of the IP was about 80%wt. of isotactic polypropylene and 20%wt, of ethylene/propylene rubber (EPR). The EPR typically contains about 20-30%wt. of ethylene, which yields a rubber with essentially zero crystallinity. The IP/HDPE blend compositions are presented in Table 1, along with the observed mechanical property data.

The IP/HDPE blends exhibited a monotonic decrease in stiffness (flexural modulus) and a maximum in impact strength (notched Izod) with increasing HDPE content, as shown in Figures 3 and 4. The mechanical properties behavior of these blends was taught in a patent[4], however, the morphological and micromechanical explanation of the phenomenon have not been provided.

The AFM of the virgin IP in Figure 5 shows typical morphology with a minor fraction of a dispersed ethylene/propylene rubber (EPR) phase (dark, dispersed particle domains) dispersed in a continuous polypropylene phase (bright continuous region). The EPR is thermodynamically incompatible with the isotactic PP and, therefore, the PP and EPR form a two-phase system. The EPR particles have an included sparse population of crystalline polyethylene matrix phase (bright region inside dark EPR particle domains), which is typical of reactor-grade IP's.[3] The crystalline polyethylene is thermodynamically incompatible with the polypropylene and the EPR, but prefers to reside as a separate phase inside the EPR domains.

The addition of up to about 10%wt. HDPE in the IP resulted in the EPR domains becoming volume-filled with HDPE, as shown in Figure 6. This system resulted in the highest impact strength, accompanied by only a modest decrease in stiffness (see Table 1). At higher loadings of HDPE the capacity of the EPR domains to contain the HDPE was exceeded and this resulted in a shift of the phase structure toward a three-phase system in which the HDPE formed a separate, third phase, within the PP matrix phase. These systems exhibited decreased impact strength relative to 10%wt. HDPE loading. As loading of HDPE increased above 10%wt., the progressive disappearance of the HDPE from inside the EPR domains indicated this shift from a two-phase to a three-phase system. At a loading of 40%wt. HDPE, the essentially complete removal of the crystalline polyethylene phase from the dispersed EPR particles may be observed in Figure 7. Since the crystalline polyethylene material is absent from

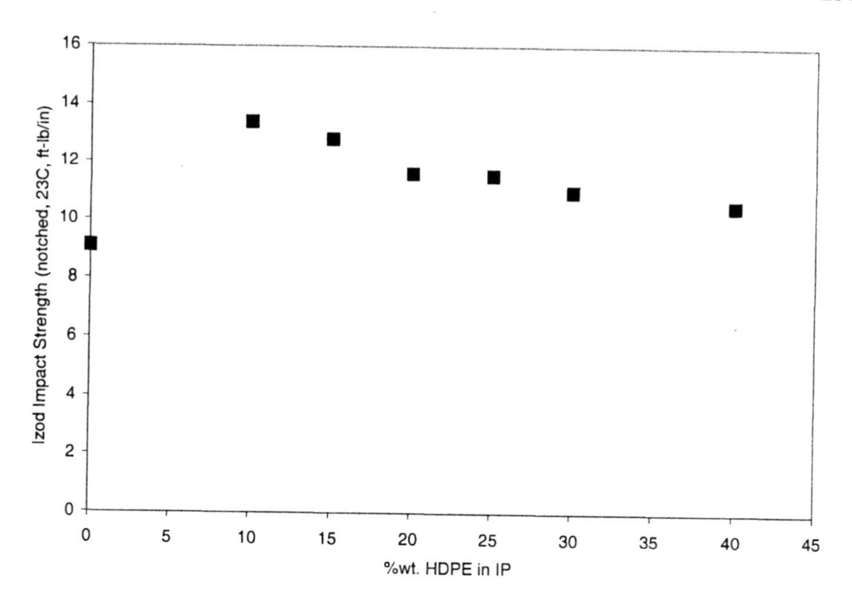

Figure 3 – Flexural modulus versus HDPE content in IP/HDPE blends.

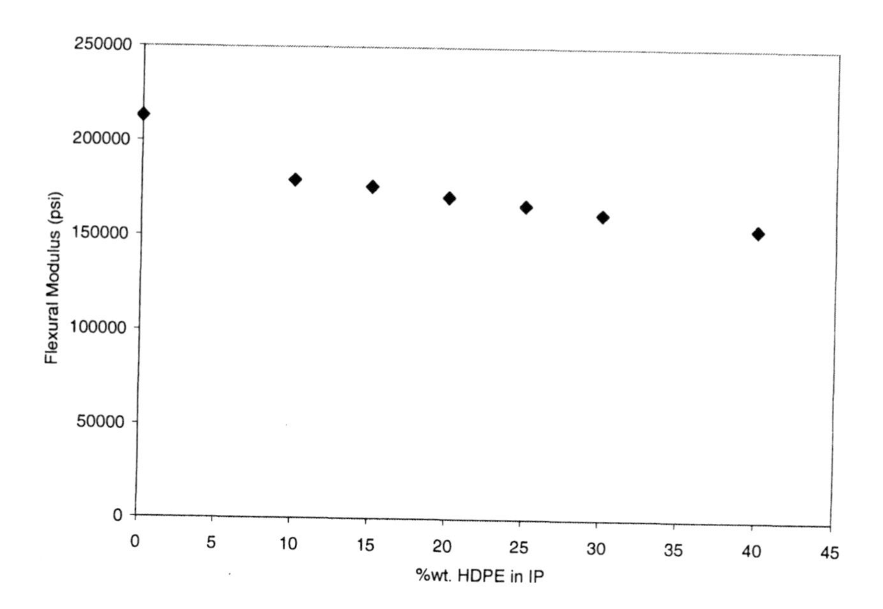

Figure 4 – Notched Izod impact strength versus HDPE content in IP/HDPE blends.

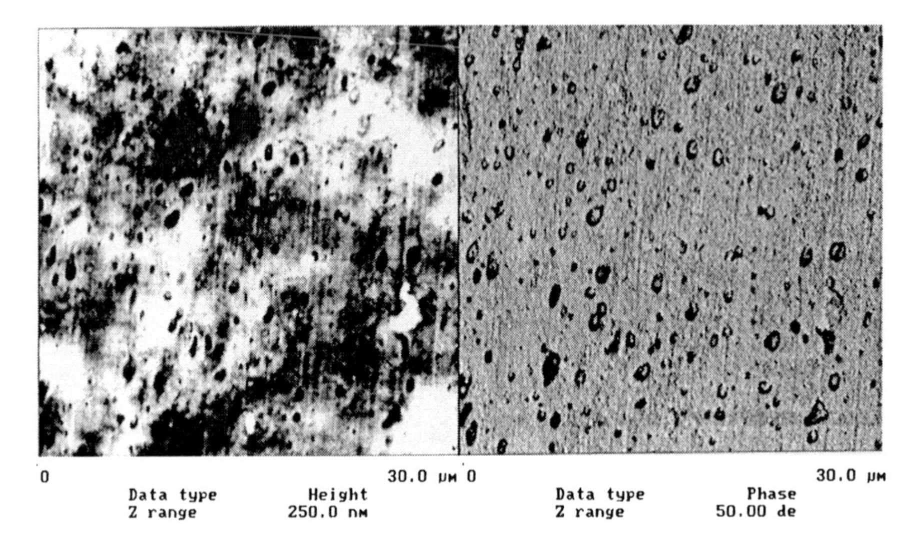

Figure 5 – AFM topographic (left image) and phase contrast (right image) photomicrographs of the virgin IP. The sample was compression molded and a flat face microtomed for AFM observation.

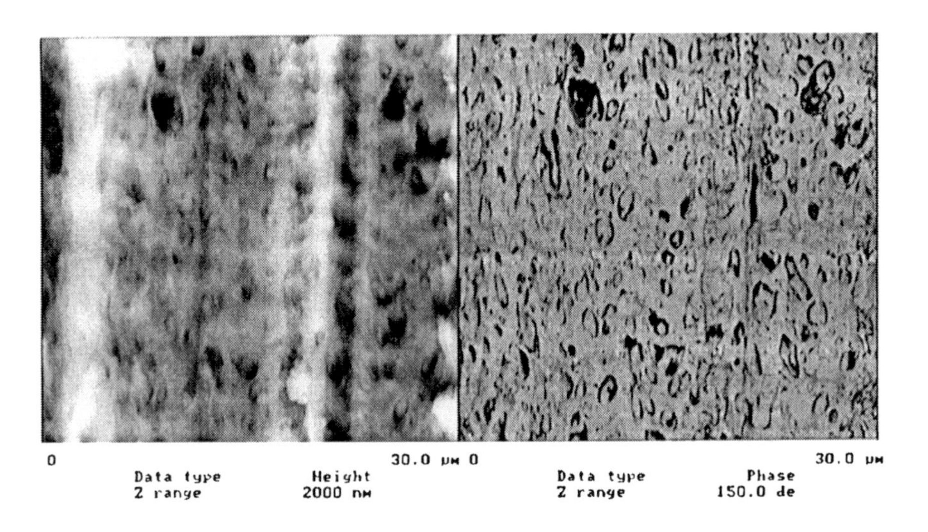

Figure 6 – AFM topographic (left image) and phase contrast (right image) photomicrographs of the 10%wtHDPE/90%wt. IP blend. The sample was compression molded and a flat face microtomed for AFM observation.

the EPR domains, it is apparent that the HDPE has formed a separate, third phase. The HDPE phase apparently coexists with the polypropylene matrix phase, however, since the polyethylene phase is hard and crystalline, as is the polypropylene matrix phase, the third, polyethylene phase is not apparent in the phase image of Figure 7.

The presence of the HDPE in the EPR particles results in an increase in impact strength and a decrease in stiffness. The addition of the lower modulus HDPE is clearly responsible for the decrease in stiffness. The increased impact strength is due to more complex factors, some of which include:

- The HDPE inside the EPR particles provides a rigid frame-work (i.e. produces an anchoring effect) that aids in energy absorption around the rubber particles.

- The presence of the HDPE inside the rubber particles increases the interfacial surface area of the EPR/PP-matrix interface, thereby extending the rubber's capability to transfer stresses to the matrix, resulting in greater energy absorption (i.e. higher impact strength).

Table 1 – IP/HDPE Blend Compositions and Mechanical Properties

IP wt.%	*HDPE wt.%*	*Flexural Modulus (1% secant) (psi)*	*Notched Izod @ 23°C (ft-lb/in)*
100	0	212900	9.09
90	10	179300	13.39
85	15	176100	12.81
80	20	170800	11.60
75	25	166600	11.54
70	30	162100	10.99
60	40	154700	10.50

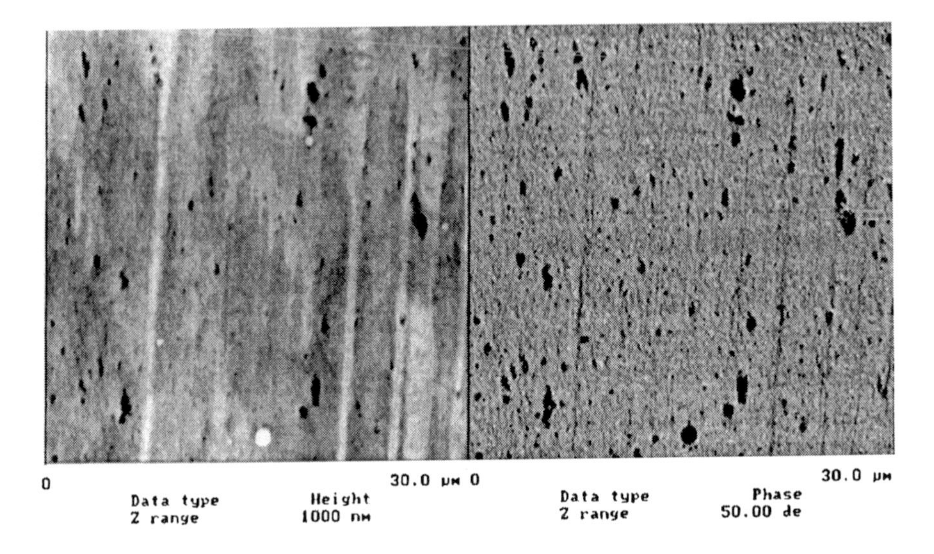

Figure 7 – AFM topographic (left image) and phase contrast (right image) photomicrographs of the 40%wtHDPE/60%wt. IP blend. The sample was compression molded and a flat face microtomed for AFM observation.

TPO/Clay Nanocomposites

In a study of polymer-layered silicate nanocomposite (PLSN) formation in a polyolefin, AFM was employed as a characterization technique to probe the morphology of the resulting PLSN systems.[5] Thermoplastic olefin (TPO)/Clay nanocomposites were made at clay loadings varying from 0.6 to 6.7 wt%. The TPO was a commercial thermoplastic olefin with 9.5 MFR (melt flow rate) and 103,500 psi flexural modulus. The TPO was composed of about 70%wt. of isotactic polypropylene and about 30%wt. of ethylene/propylene rubber (EPR). The EPR typically contains about 20-30%wt. of ethylene, which yields a rubber with essentially zero crystallinity.

The nanocomposites were prepared by blending the TPO with maleic anhydride grafted PP (PP-MA, 1.0%MA content) as a compatiblizer and Cloisite® 20A natural montmorillonite clay modified with a quaternary ammonium salt

(Southern clay products). The mechanical properties (flexural modulus and impact strength) were measured on the TPO/clay nanocomposite systems.

These TPO/clay systems largely exhibited dispersion of the clay into stacks of about 10-50 nm in thickness, corresponding to stacks of about 5-25 clay platelets, as determined by TEM. (A small fraction of clay platelets were exfoliated into essentially single platelet arrays.) These stacks are often called tactoids and exhibit clay spacing (clay long period) of about 2-3 nm, due to intercalation of the PP-MA into the clay galleries.[6] The unmodified clay spacing was ~1 nm).[6] Clay long period was measured by x-ray diffraction (XRD).

The polymer morphology of these TPO/clay nanocomposites was investigated with AFM. The EPR particle morphology in the TPO was found to undergo progressive particle break-up and decrease in particle size, as clay loading increased in the range from 0.6 to 5.6%wt. clay, as shown in the AFM photomicrographs in Figure 8. The decrease in measured average particle diameter is shown in Figure 9. The breakup of the EPR particles was suspected to be due to the increasing melt viscosity observed as clay loading increased, and/or the accompanying chemical modifiers on the clay, acting as interfacial agents, reducing the interfacial tension with concomitant reduction in particle size. The progressive modification of the polymer morphology was used as part of the argument to infer structure/mechanical property relationships in the nanocomposite systems.[5]

Polymer Characterization by AFM

The examples given of polymer characterization by AFM in this chapter were extracted from more comprehensive studies of polymer structure/property relationships. In many studies of this kind AFM contributed partial, but often critical, information. Studies in which AFM contributed a critical component of the information continue to increase in the polymer literature.[2] Comprehensive reviews of AFM applications to polymer characterization contain numerous references to specific studies of various kinds.[1, 7-12]

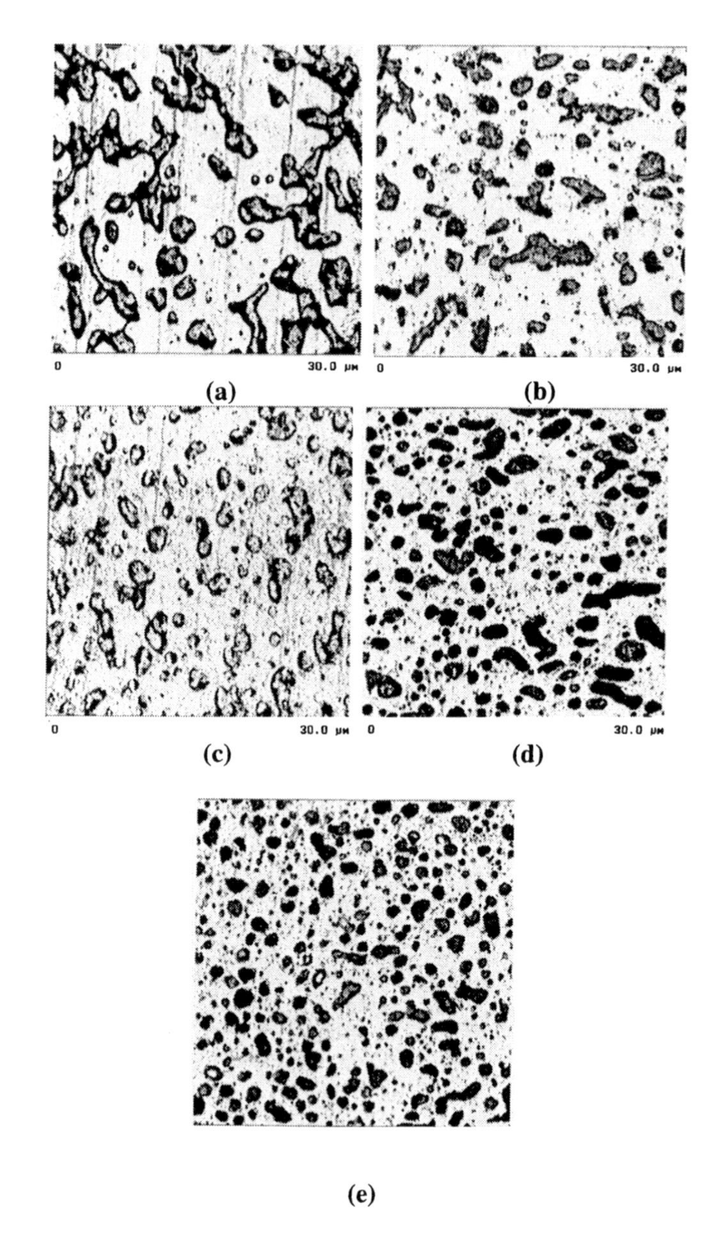

Figure 8 - AFM phase contrast images of (a) TPO-0 (0wt% clay), (b) TPO-1 (0.6wt% clay), (c) TPO-3 (2.3wt% clay), (d) TPO-4 (3.3wt% clay), and (e) TPO-6 (5.6wt% clay). Reproduced with permission from reference 5. Copyright 2004 Wiley Periodicals, Inc.

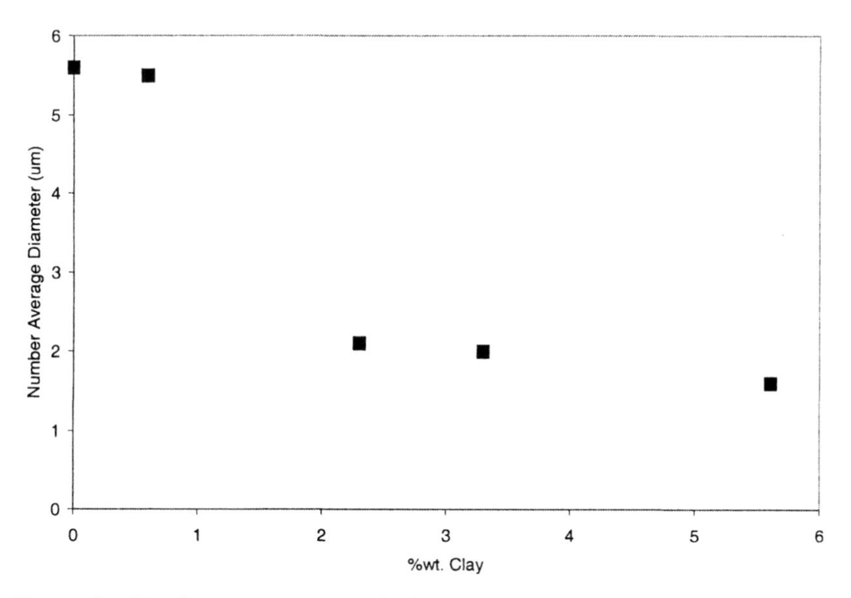

Figure 9 – Number-average particle diameter versus clay loading in TPO/Clay nanocomposites.

Conclusions

Atomic force microscopy (AFM) has distinguished itself as a central technique for the characterization of polymers. It has been especially effective in the characterization of multi-phase polymer systems. One aspect of this technique, in comparison to the electron microscopy techniques, is the ease of sample preparation. AFM requires little or no sample preparation and preserves sample structure, whereas SEM and TEM , typically, require much more sample preparation, which often destroys or modifies sample structure in the process. AFM has the attribute of directness of observation and, therefore, reveals structural features of natural surfaces or cross-sections of fabricated polymer articles, which are often difficult to observe by the electron microscopy techniques, due to the necessity of more extensive sample preparation.

Acknowledgements

The author gratefully acknowledges helpful discussions with Dr. Sameer Mehta and Chris Dickerson, and the technical assistance in sample preparation of Dr.Ayush Bafna, Amy Weiskittel and Brenda Racke.

References

1. D.A. Chernov, S. Maganov, Atomic Force Microscopy, Ch. 19 in *Comprehensive Desk Reference of Polymer Characterization and Analysis*, R.F. Brady, Jr. Ed., Oxford University Press, Oxford, 2003.
2. F.M. Mirabella, Polymer News in press.
3. F.M. Mirabella, Polymer, **1993**, 34, 1729.
4. James F. Ross, "High Impact Visbroken Polymeric Blends", US Patent 4,375,531, March 1, 1983.
5. S. Mehta, F.M. Mirabella, K. Rufener, A. Bafna, J. Appl. Polym. Sci., **2004**, 92, 928.
6. F.M. Mirabella, "Polypropylene/Clay and Thermoplastic Olefin/Clay Nanocomposites" in *Encyclopedia of Nanoscience and Nanotechnology* ed. Hari Singh Nalwa, American Scientific Publishers, 2004.
7. S.S. Sheiko, Adv. Polym. Sci., **2000**, 151, 61.
8. S.N. Magonov, Reneker, D.H., **1997**, Annu. Rev. Mater. Sci., 27, 175.
9. V.V. Tsukruk, Rubber Chem. Technol., **1997**, 70, 430.
10. B.D. Ratner, Tsukruk, (eds.), *Scanning Probe Microscopy of Polymers*, ACS Symp. Series 694, 1998, American Chemical Society, Washington, DC.
11. K.D. Jandt, Mater. Sci. Eng., **1998**, R21, 221.
12. T. Kajiyama, K. Tanaka, S-R. Ge, A.T. Takahara, Prog. Surf. Sci., **1996**, 52, 1.

Chapter 17

Evaluation of Surface Composition in Miscible Polymer Blends by Lateral Force Microscopy

Keiji Tanaka[1], Atsushi Takahara[2], and Tisato Kajiyama[3]

[1]Department of Applied Chemistry, Faculty of Engineering and [2]Institute for Materials Chemistry and Engineering, Kyusha University, Hakozaki, Higashi-ku, Fukuoka 812–8581, Japan
[3]President, Kyushu University, Hakozaki, Higashi-ku, Fukuoka 812-8581, Japan

A novel method to characterize surface composition in blends of two chemically identical polymers with different molecular masses using lateral force microscopy is proposed. Extending Gordon-Taylor equation to surface, surface composition in the blends would be obtained by measuring surface glass transition temperature of each constituent as well as the blend. Surface composition in blends of polystyrene (PS) and deuterated PS (dPS) obtained by this method was in good accordance with the result by well-established surface spectroscopy. Finally, surface composition in blend films of two PSs with different molecular masses was experimentally elucidated. The surface enrichment of a smaller mass component became more remarkable with increasing molecular mass disparity between the two components.

Introduction

In general, synthetic polymers have a broad molecular weight distribution, and the surface of a film prepared by them would be covered with smaller molecular weight components. Hence, it is of importance to clarify an effect of shorter chains on aggregation states and physical properties at the surface. One of easier ways to study experimentally such is to adopt a model blend system composed of two monodisperse polymers with different molecular measses, namely, a polymer with bimodal molecular weight distribution. To date, we have studied on surface relaxation behavior of polystyrene (PS) films with bimodal molecular weight.[1] Surface relaxation in the films was dominated by the smaller molecular mass component rather than the higher one. Although it was clear that the smaller molecular weight component was preferentially partitioned to the surface, it was impossible only from this experiment to deduce to what extent the shorter component were present at the surface. The objective of this study is to propose a novel experimental method to study surface composition in blends of chemically identical two polymers based on our surface glass transition temperature (T_g^s) measurements,[2] which do not need any labeling procedures, using lateral force microscopy (LFM).

Experimental

Materials and Film Preparation

Monodisperse PSs with various number-average molecular masses (M_n) were synthesized by a living anionic polymerization using *sec*-butyllithium as an initiator and methanol as a terminator. Also, monodisperse PSs with M_n of 981k and 1.46M, and monodisperse dPS with M_n of 847k were purchased from Polymer Laboratories Co, Ltd. Blend films were spin-coated from toluene solutions onto cleaned silicon wafers with a native oxide layer. The films were dried at 296 K for more than 24 h and then annealed at 423 K for, at least, 48 h under vacuum. Each film thickness was approximately 200 nm or even thicker.

Surface Characterizations

T_g^s of homo and blend films was determined on the basis of surface relaxation behavior using LFM (SPA 300 HV, Seiko Instruments Industry Co., Ltd.) with an SPI 3800 controller. LFM measurement was carried out at various temperatures in vacuo. A cantilever with the bending spring constant of

0.11±0.02 N m^{-1}, of which both sides were uncoated or coated by gold, was used. Applied force to the cantilever was set to be 10 nN in a repulsive force region. Bulk glass transition temperature (T_g^b) of the homo and blend samples was measured by differential scanning calorimetry (DSC8230, Rigaku Co., Ltd.).

Surface composition in blend films composed of PS and dPS was examined by time-of-flight secondary ion mass spectroscopy (ToF-SIMS). A Physical Electronics TRIFT-II ToF-SIMS instrument was used. Positive spectra were obtained with 15 kV primary pulsed Ga^+ ion beam (pulse width of 13 ns) with 2 nA beam current. The scan area was 100 μm x 100 μm.

Results and Discussion

Surface Composition in (PS/dPS) Blend Films by Surface Spectroscopy

At first, it was confirmed by surface spectroscopy how surface composition in (PS19.7k/dPS847k) blend films differs from the corresponding bulk one. This blend system was completely miscible at the range of room temperature to annealing temperature of 423 K.[3] Under static conditions (with a maximal ion dose of 10^{12} ions cm^{-2}), information provided by ToF-SIMS is obtained from a monolayer depth region. Figure 1 shows the typical ToF-SIMS spectrum of the (PS19.7k/dPS847k) film with the bulk composition of (41.5/58.5) in volume.[4] The intense peaks observed at 91.05 and 98.10 amu can be assigned to tropylium ion, $C_7H_7^+$, and its deuterated species, $C_7D_7^+$, respectively. Also, the peaks at 92.06 and 97.09 amu are due probably to $C_7DH_6^+$ or $C_6C^{13}H_7^+$, and $C_7D_6H^+$. While the secondary ion intensities at 91.05 and 98.10 amu monotonically

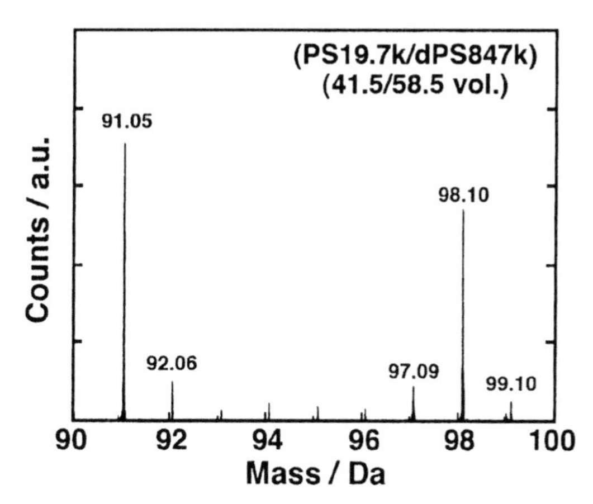

Figure 1. Positive secondary ion time-of-flight mass spectrum of (PS19.7k/dPS847k) blend film with the bulk PS fraction of 41.5 vol%.

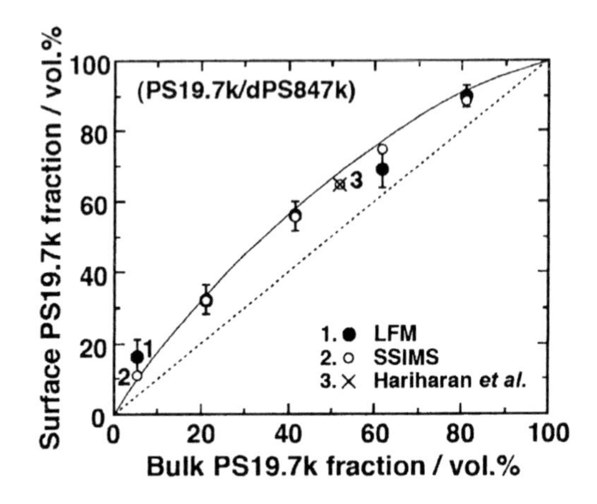

Figure 2. Relation of surface hPS fraction to bulk composition in (PS19.7k/dPS847k) blend films. Surface composition was examined by two different techniques; ToF-SIMS and LFM. Data by Hariharan et al. using NR was adopted from Table 1 in reference 10. The broken line denotes the case where the surface and bulk compositions are the same. The solid curve is drawn to guide the eye.

increased and decreased with increasing feed fraction of PS into the blend, respectively, those at 92.06 and 97.09 amu were not necessarily proportional to the blend ratio. Hence, the surface PS fraction in the blend was estimated from the value of I_{91} / (I_{91}+I_{98}), where I_i was the secondary ion intensity at i amu, provided that a contribution of chain end fragments to I_{91} was corrected by following the procedure of Vanden Eynde et al.[5] Also, the (I_{91}+I_{98}) / (I_{91}+I_{92}+I_{97}+I_{98}) value was accordance with the value of I_{91} / (I_{91}+I_{98}) within 5%. The surface PS fraction in the (PS19.7k/dPS847k) blend films so obtained was plotted in Figure 2. The surface fraction of PS19.7k was higher than the bulk one at all blend ratios employed. In general, the surface of (PS/dPS) blends, in which both components have the comparable high M_n, is covered with dPS owing to its lower surface energy. However, PS19.7k, which was the lower M_n component, was enriched at the blend surface in this study. This result can be understood by considering a situation that the PS19.7k suffers less of the conformational entropic penalty at the surface in comparison with dPS847k.[6]

Surface Composition in (PS/dPS) Blend Films by Lateral Force Microscopy

In general, T_g^b of a miscible binary polymer blend is well expressed by T_g^b of each constituent and the blend ratio in volume. This means that the bulk composition in the belnd can be deduced by T_g^b of each constituent and the blend.

This notion is based on the additivity rule of the free volume between the components, and is widely accepted as Gordon-Taylor equation.[8] Although the free volume of a component at the surface would be not the same as that in the interior bulk region, such a situation is common for both components. Hence, invoking that the additivity rule of the free volume is held even at the surface, Gordon-Taylor equation is applied to T_g^s. Since we are able to examine T_g^s using LFM,[2] surface composition in blends composed of two polymers can be obtained by

$$\phi_1 = (T_{g,2}^s - T_{g,mix}^s) / (T_{g,2}^s - T_{g,1}^s) \qquad (1)$$

To confirm the validity of the aforementioned strategy, surface composition in the (PS19.7k/dPS847k) blend films is examined. Figure 3 shows the temperature dependence of lateral force at the scanning rate of 1 μm s^{-1} for the PS19.7k/dPS847k) blend films with various bulk blend ratios. Since lateral force is originated from energy dissipation during a tip slides on the sample surface, the temperature dependence of lateral force corresponds well to the dynamic loss modulus variation with temperature. Peaks observed on the temperature-lateral force curves are assigned to the α_a-absorption corresponding to the segmental motion.[2] Hence, an onset temperature on the curve, that is, the

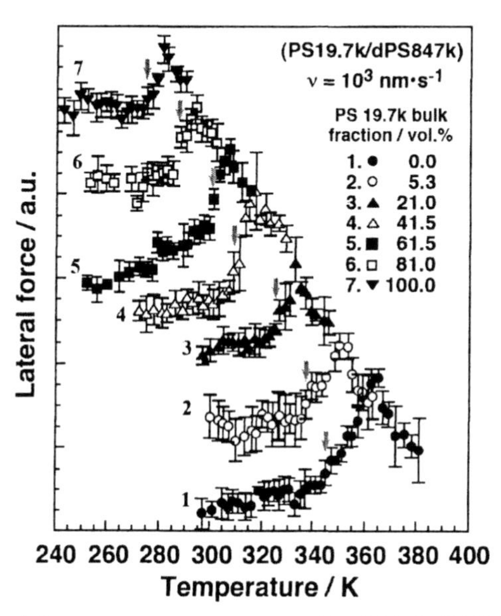

Figure 3. Lateral force-temperature curves for (PS19.7k/dPS847k) blend films with various compositions at the scanning rate of 1 μm s^{-1}. (Reproduced with permission from reference 7. Copyright 2002 American Chemical Society).

temperature at which lateral force starts to increase, can be empirically defined as T_g^s,[2] as marked by arrows in Figure 3. T_g^s of the PS19.7k and dPS847k films were evaluated to be 279 and 347 K, respectively, being much lower than the corresponding T_g^b values. End groups of the polymers used have a lower surface free energy in comparison with the main chain part, and thus, are preferentially segregated at the surface.[2,9] Since the end groups have a larger freedom compared with the main chain part, an excess free volume is supposed to be induced at the surface, resulting in the enhanced chain mobility at the surface. In addition, there exists the free space on the polymer surface. This makes the cooperative movement for the surface segmental motion easier.[2] Hence, T_g^s was much lower than the corresponding T_g^b. Besides, T_g^s value in the blend films was dependent on the blend ratio.

Figure 4 shows T_g^b and T_g^s of the (PS19.7k/dPS847k) films with various blend compositions. The dot lines denote the additivity rule of glass transition temperature for bulk and surface. T_g^b linearly decreased with increasing PS19.7k fraction in the blend. In contrast, the relation between T_g^s and blend ratio was not linear but negatively deviated, as shown in Figure 4. This is because the abscissa is expressed by the bulk blend ratio, although PS19.7k should be preferentially segregated at the blend surface. The surface fraction of the PS19.7k in the (PS19.7k/dPS847k) blend films based on the T_g^s measurements by LFM was plotted in Figure 2 as filled circles. In addition, surface composition in similar blend films by Hariharan *et al.* using neutron reflectivity (NR) was shown in it.[10] Surface composition among three independent

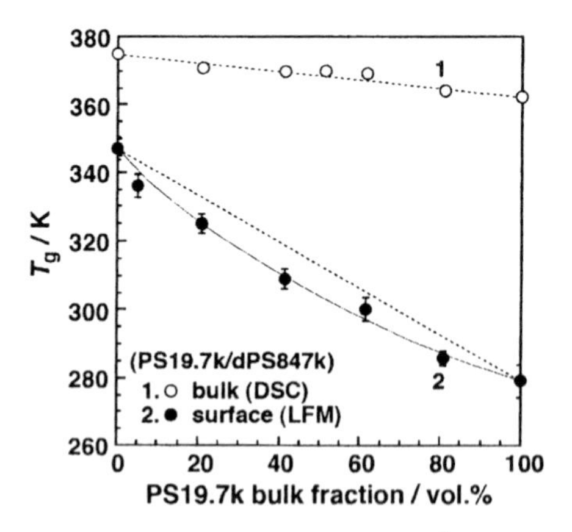

Figure 4. Bulk blend ratio dependences of T_g^b by DSC and T_g^s by LFM. Each dot line denotes a simple additivity rule of T_g. The solid curve for T_g^s is drawn to guide the eye. (Reproduced with permission from reference 7. Copyright 2002 American Chemical Society).

experiments such as ToF-SIMS, LFM and NR are in excellent agreement with one another. Hence, it can be claimed that the T_g^s measurement enables us to gain direct access to surface composition in miscible binary blend films. Here, two points should be emphasized. The first point is that T_g^s based on our LFM measurement is extremely reliable. Otherwise, there is no reason why the surface composition via the T_g^s measurement is in accordance with that by surface spectroscopy. The second is the following. In all of well-established surface characterization techniques, one component must be deuterated to confer a contrast between components. However, our method using LFM is not the case, meaning that surface composition in blends composed of two chemically identical polymers can be experimentally obtained.

Surface Composition in (PS/PS) Blend Films by Lateral Force Microscopy

We now turn to the surface segregation in blend film of low molecular weight PS (LMW-PS) and PS with M_n of 1.46M. For a comparison, (LMW-PS/dPS847k) system was also examined. LMW components for the blends were successively changed in terms of M_n. Figure 5 shows surface composition in the (LMW-PS/PS1.46M) and the (LMW-PS/dPS847k) blends as a function of M_n for the LMW component. The (LMW-PS/PS1.46M) blend corresponds to PS with bimodal molecular weight distribution. The bulk blend ratios of the (LMW-PS/PS1.46M) and the (LMW-PS/dPS847k) were 50 vol% and 51.6 vol% (=50 wt%), respectively. Surface composition in the (LMW-PS/PS1.46M) films was obtained by T_g^s measurements via eq. (1). Even in the case of the (PS140k/PS1.46M), the distinct partition of the PS140k to the surface was observed. The extent of the surface concentration of the LMW-PS became remarkable with decreasing M_n of the LMW-PS, as shown by filled circles in Figure 5. Although the surface segregation of the LMW component in chemically identical binary blend films has been theoretically predicted by Hariharan *et al.* based on a compressible mean-field lattice model,[11] the experimental evidence has not been reported thus far. Hence, this is the first experimental report to prove quantitatively the surface segregation of the LMW component. In Figure 5, surface composition of the (LMW-PS/dPS847k) films was shown and compared with the reported results for the (LMW-PS/dPS571k) blends by Hariharan *et al.* using neutron reflection.[10] They were in good agreement with each other. In the case of the blends composed of LMW-PS and high molecular weight (HMW) dPS, which component is enriched at the surface is decided by the competition between energetic and entropic effects. For example, in the case of symmetric (PS/dPS) blends, the energetic effect based on the discrepancy of the polarizability between C-H and C-D bonds dominates, resulting in the surface enrichment of dPS. On the other hand, PS is partitioned

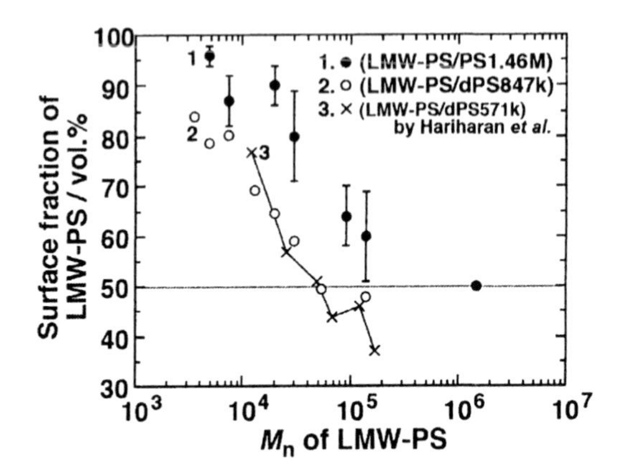

Figure 5. Surface compositions in (LMW-PS/PS1.46M) and (LMW-PS/dPS847k) blend films as a function of M_n for LMW-PS. For a comparison, the data for (LMW-PS/dPS571k) blends by Hariharan et al. is plotted from Ref. 10. (Reproduced with permission from reference 7. Copyright 2002 American Chemical Society).

to the surface if PS is much shorter than dPS. In general, a polymer chain present at the surface might have a flattened conformation.[6] This means that the number of possible conformations for a chain at the surface would decrease in comparison with the case where the chain has a random coil conformation in the interior bulk region. Consequently, shorter chains suffer less of an entropic penalty than longer chains, and thus, are enriched at the surface to minimize the free energy of the system.

Conclusions

An experimental method to characterize the surface concentration in miscible binary polymer mixtures, which is based on our T_g^s measurements using LFM, is proposed. Surface composition in PS/dPS blends so obtained by LFM was in good accordance with the results by surface spectroscopy. Since our method does not need any labeling procedure unlike usual spectroscopic techniques, the surface composition of the LMW component in the PS films with various binary molecular weight distributions was experimentally clarified.

Acknowledgement

We are most grateful for fruitful discussion with Prof. Toshihiko Nagamura, Kyushu University. This was in part supported by a Grant-in-Aid for Scientific Research (A) (#13355034) from the Japanese Society of Promotion of Sciences and Project on Nanostructured Polymeric Materials from NEDO (New Energy and Industrial Technology Development Organization) .

References

1. Tanaka, K.; Takahara, A.; Kajiyama, T. *Macromolecules* **1997**, 30, 6626.
2. Tanaka, K.; Takahara, A.; Kajiyama, T. *Macromolecules* **2000**, 33, 7588.
3. Bates, F. S.; Wignall, G. D. *Phys. Rev. Lett.* **1986**, *57*, 1429.
4. Takahara, A; Kawaguchi, D; Tanaka, K.; Tozu, M.; Hoshi, T.; Kajiyama, T. *Appl. Surf. Aci.* **2003**, *203-204*, 538.
5. Vanden Eynde, X.; Bertrand, P.; Jérôme, R. *Macromolecules* **1997**, *30*, 6407.
6. Bitsanis, I. A.; Brinke, G. *J. Chem. Phys.* **1993**, *99*, 3100.
7. Tanaka, K.; Takahara, A.; Kajiyama, T.; Tasaki, S. *Macromolecules* **2002,** 35, 4702.
8. Gordon, M.; Taylor, J. S. *J. Appl. Chem.* **1952**, *2*, 493.
9. Tanaka, K.; Taura, A.; Ge, S.-R.; Takahara, A.; Kajiyama, T. *Macromolecules* **1996**, *29*, 3040.
10. Hariharan, A.; Kumar, S. K.; Russell, T. P. *J. Chem. Phys.* **1993**, *98*, 4163.
11. Hariharan, A.; Kumar, S. K.; Russell, T. P. *J. Chem. Phys.* **1993**, *99*, 4041.

Indexes

Author Index

Subject Index

B

C

D

E

F

G

H

I

J

K

L

M

N

O

P

Q

R

S

U

W

X

Y